Lecture Notes of the Institute
for Computer Sciences, Social Informatics
and Telecommunications Engineering   94

W0227345

Ramjee Prasad   Károly Farkas
Andreas U. Schmidt   Antonio Lioy
Giovanni Russello   Flaminia L. Luccio (Eds.)

Security and Privacy in Mobile Information and Communication Systems

Third International ICST Conference
MobiSec 2011
Aalborg, Denmark, May 17-19, 2011
Revised Selected Papers

 Springer

Volume Editors

Ramjee Prasad
Aalborg University, 9220 Aalborg, Denmark
E-mail: prasad@es.aau.dk

Károly Farkas
University of West Hungary, Sopron
and Budapest University of Technology and Economics
9400 Budapest, Hungary
E-mail: farkas@inf.nyme.hu

Andreas U. Schmidt
Novalyst IT AG, 61184 Karben, Germany
E-mail: andreas.schmidt@novalyst.de

Antonio Lioy
Politecnico di Torino, 10129 Torino, Italy
E-mail: lioy@polito.it

Giovanni Russello
CREATE-NET Research Consortium, 38121 Trento, Italy
E-mail: giovanni.russello@create-net.org

Flaminia L. Luccio
Università Ca' Foscari di Venezia, 30172 Mestre (VE) Italy
E-mail: luccio@unive.it

ISSN 1867-8211   e-ISSN 1867-822X
ISBN 978-3-642-30243-5   e-ISBN 978-3-642-30244-2
DOI 10.1007/978-3-642-30244-2

Springer Heidelberg Dordrecht London New York

Library of Congress Control Number: 2012937592

CR Subject Classification (1998): I.2, C.2, H.4, H.3, K.4.4, K.6.5, D.4.6, J.1

Typesetting: Camera-ready by author, data conversion by Scientific Publishing Services, Chennai, India

Printed on acid-free paper

Springer is part of Springer Science+Business Media (www.springer.com)

Preface

MobiSec 2011 was the third ICST conference on security and privacy for mobile information and communication systems. This year, MobiSec was co-organized together with the Annual Workshop of the Center for TeleInFrastruktur (CTIF) of Aalborg University. The beautiful city of Aalborg, in North Jutland, Denmark, is a perfect setting to carry out networking and discussions on the various mobile security topics brought forward by the conference speakers and participants. The technical program featured more than 15 technical papers from MobiSec and 19 presentations from the CTIF Annual Workshop. CTIF is a global center with branches in Denmark, Italy, India, Japan, and the USA conducting research in wireless technologies with special focus on security and mobility aspects. During this year's CTIF Annual Workshop, attendees included colleagues from Denmark, Noway, Greece, China, India, and the USA and the topics spanned DDoS in MANETs, secure cognitive radio networks along with novel mechanisms for RRM in next-generation networks and novel social networking applications.

MobiSec 2011 covered some of the most active areas of research in mobile security with its three focus areas: Machine-to-Machine Communication Security, Policies for Mobile Environments, and Mobile User Authentication and Authorization. With the orientation toward applications, MobiSec is a perfect interface between academia and industry in the field of mobile communications. This is corroborated by our many prominent guests from all over the world.

This second edition of MobiSec, beyond attracting excellent scientific papers, featured a very interesting by-program. Our distinguished keynote speakers from important organizations such as the Tata Consulting Group and Beijing University of Posts and Telecommunications, Rajiv Gandhi University of Knowledge Technologies (RGUKT), Hyderabad, India, provided essential insight into the latest trends in industry and academia to set a framework for the theme of mobile security.

Within the vast area of mobile technology research and application, this second MobiSec strove to make a small, but unique, contribution. Our aim was to build a bridge between top-level research and large-scale applications of novel kinds of information security for mobile devices and communication. It was a privilege to serve this event as the Organizing committee.

Many people contributed to the organization of MobiSec. It was a privilege to work with these dedicated persons, and we would like to thank them all for their efforts. The Organizing Committee as a whole created a frictionless, collaborative work atmosphere, which made our task an easy one. Special thanks must go to the local Organizing Committee Chair Neeli Prasad and Rasmus Hjorth Nielsen from the Center for TeleInfrastruktur, at Aalborg University. A high-quality conference cannot be created without the help of the distinguished members of the Program Committee, overseen by the TPC Chairs Giovanni

Russello, Antonio Lioy, and Flaminia Luccio. Thanks to all other members of
the Organizing Committee: Shiguo Lian (Publications Chair), Vincent Naessens
(Workshops Chair), Dirk Kröselberg (Panels and Keynotes Chair). The support
of the conference organizer, ICST, represented by Aza Swedin and Anna Sterzi of
the European Alliance for Innovation, is greatly acknowledged. Finally, we would
like to thank Imrich Chlamtac and the members of the Steering Committee for
their support and guidance during these months.

Concluding, we hope the reader will find this proceedings volume of MobiSec 2011 stimulating and thought-provoking. We encourage You to join us at
MobiSec 2012 in Frankfurt, Germany.

Ramjee Prasad
Andreas U. Schmidt
Antonio Lioy
Shiguo Lian

Organization

General Chair

Ramjee Prasad   Aalborg University, Denmark

General Co-chairs

Károly Farkas   University of West Hungary, Sopron, and
Budapest University of Technology and
Economics, Budapest, Hungary

Andreas U. Schmidt   Novalyst IT AG, Karben, Germany

Steering Committee Chairs

Imrich Chlamtac (Chair)   President, CREATE-NET Research
Consortium, Trento, Italy

Andreas U. Schmidt   Director, Novalyst IT AG, Karben, Germany

Technical Program Co-chairs

Antonio Lioy   Politecnico di Torino, Italy

Giovanni Russello   CREATE-NET Research Consortium, Trento,
Italy

Flaminia Luccio   University Ca'Foscari, Venice, Italy

Local Program Chair

Neeli R. Prasad   Aalborg University, Denmark

Publications Chair

Shiguo Lian   France Telecom R&D, Beijing, China

Workshops Chair

Vincent Naessens   KaHo Sint-Lieven, Gent, Belgium

Panels and Keynotes Chair

Dirk Kröselberg   Siemens CERT, Munich, Germany

Web Chair

Rasmus H. Nielsen   Aalborg University, Denmark

Conference Coordinator

Aza Swedin   European Alliance for Innovation
Anna Sterzi   European Alliance for Innovation

Technical Program Committee

Selim Aissi   Intel, USA
Claudio A. Ardagna   Università di Milano, Italy
Lejla Batina   Katholieke Universiteit Leuven, Belgium
Francesco Bergadano   Università degli Studi di Torino, Italy
Reinhardt A. Botha   Nelson Mandela Metropolitan University,
                             South Africa
Marco Casassa-Mont   HP Labs, UK
Rocky K. C. Chang   Hong Kong Polytechnic University, China
Changyu Dong   Imperial College, UK
Sara Foresti   Università di Milano, Italy
Ashish Gehani   SRI International, USA
Vaibhav Gowadia   Imperial College, UK
Marco Hauri   ASCOM, Switzerland
Kazukuni Kobara   AIST, Japan
Geir Myrdahl Koien   University of Adger, Norway
Andreas Leicher   Novalyst IT, Germany
Jiqiang Lu   Eindhoven University of Technology,
                             The Netherlands
Flaminia Luccio   Università Ca'Foscari, Italy
Khamish Malhotra   University of Glamorgan, UK
Fabio Martinelli   CNR Pisa, Italy
Gregorio Martinez Perez   University of Murcia, Spain
Raphael C.-W. Phan   Loughborough University, UK
Anand Prasad   NEC Laboratories, Japan
Reijo Savola   VTT, Finland
Hans Georg Schaathun   Surrey University, UK
Stefan Tillich   Graz University of Technology, Austria
Allan Tomlinson   Royal Holloway, University of London, UK
Xin Wang   ContentGuard, Inc., USA
Martin Werner   LMU Munich, Germany
Grzegorz Sowa   Comarch, Poland

Table of Contents

Conference Papers

Android Market Analysis
with Activation Patterns

Peter Teufl, Stefan Kraxberger, Clemens Orthacker, Günther Lackner,
Michael Gissing, Alexander Marsalek,
Johannes Leibetseder, and Oliver Prevenhueber

University of Technology Graz, Institute for Applied Information Processing and
Communications, Graz, Austria
{peter.teufl,stefan.kraxberger,clemens.orthacker,
guenther.lackner}@iaik.tugraz.at

Abstract. The increasing market share of the Android platform is
partly caused by a growing number of applications (apps) available on
the Android market: by now (January 2011) roughly 200.000. This popularity in combination with the lax market approval process attracts
the injection of malicious apps into the market. Android features a fine-
grained permission system allowing the user to review the permissions an
app requests and grant or deny access to resources prior to installation.
In this paper, we extract these security permissions along other metadata of 130.211 apps and apply a new analysis method called Activation
Patterns. Thereby, we are able to gain a new understanding of the apps
through extracting knowledge about security permissions, their relations
and possible anomalies, executing semantic search queries, finding relations between the description and the employed security permissions, or
identifying clusters of similar apps. The paper describes the employed
method and highlights its benefits in several analysis examples – e.g.
screening the market for possible malicious apps that should be further
investigated.

Keywords: Android Market, Activation Patterns, Machine Learning,
Security Permissions, Android Malware, Anomaly Detection, Semantic
Search, Unsupervised Clustering.

1 Introduction

As mobile operating systems start to spread from the classic smartphone platforms onto tablets and ultra mobile computers, they are experiencing a significant gain in market share. Consumer acceptance and therefore commercial success of a mobile operating system depends on several factors. Beside the quality
and usability of the user interface, the availability of applications (apps) may be
the most important feature demanded by the market. While traditional systems
provided a preinstalled set of apps, modern solutions like Apple's iOS, Google's
Android, RIM's Blackberry or Microsoft's Windows Phone 7 offer the possibility

R. Prasad et al. (Eds.): MOBISEC 2011, LNICST 94, pp. 1–12, 2012.

to access and install a wide variety of apps from different genres, ranging from games to powerful business appliances.

While Apple enforces tight policies on software distributed via their App Store for iOS, regarding security and content, Google emphasizes a more *open* philosophy, providing many liberties to Android developers, distributing their products via the Android Market.

Newly developed iOS apps are thoroughly examined by Apple engineers to keep the platform as secure as possible. This approach sometimes limits the developers' access to hardware resources like GPS receivers, cell phone functionalities or integrated cameras. As recent events have shown, even this strict approach has loopholes and apps not in line with the policies can find their ways onto the customer's devices[1]. In order to confine the impact of such incidents, Apple supposedly implemented a *kill switch* to deactivate installed apps on all devices[2]. Although not much details are known, from a security point of view it makes sense to use such a component.

Apps submitted to the Android Market are rudimentarily checked but the process is not as strict as it is for the App Store. Google seems to pursue a *delete afterwards* strategy if apps have been found of low quality or malicious. Android implements a similar functionality to Apple's kill switch to remotely remove apps installed on customer devices[3].

Google introduced a fine grained permission system for their Android platform, allowing developers to precisely define the necessary resources and permissions for their products. The customer can decide during the installation whether she wants to grant or deny access to these requested resources such as the address book, the GPS subsystem or telephone functionalities[4].

This user centric process is sometimes challenging and inducing customers to accept whatever is requested by the app, opening potential loopholes for attackers. In order to gain a better understanding on how permissions are used throughout the Android Market, this paper presents our analysis of publicly available metadata of 130.211 apps in the Android Market. The extracted metadata is comprised of several features that are displayed when the user opens an app for installation. Among the security permissions, which were of primary interest, we have extracted the description, download count, price and the category of each app. For the analysis, we propose a sophisticated method based on *Activation Patterns* that allows us to answer a wide range of questions such as:

- *Extract all wall papers that have a non-typical combination of security permissions - meaning they are anomalies.*

---

[1] http://news.cnet.com/8301-13579_3-10464021-37.html
[2] http://www.telegraph.co.uk/technology/3358134/
Apples-Jobs-confirms-iPhone-kill-switch.html
[3] http://www.engadget.com/2010/06/25/google-flexes-
biceps-flicks-android-remote-kill-switch-for-the/
[4] The user can accept either all permissions and install the app or reject all permission by not installing the app. Accepting or rejecting just a subset of these permissions is not possible.

- *Is the usage of security permissions different in free/payed apps?*
- *Which permissions are typical when the term "navigation" is used within the description?*
- *What are the most relevant features when analyzing popular security apps?*
- *Cluster popular apps according to their description or security permissions.*

The paper first describes the Android permission system and discusses related work, then describes the *Activation Pattern* concept and finally shows how it is applied to the metadata of 130.211 Android Market apps.

2 Android Permission Mechanism

Android's security architecture ensures the isolation of apps from each other as well as from the system. Communication and resource sharing are subject to well-defined access restrictions. Android apps are executed in their own Linux process with their own unique user- and group ID (UID), which allows for protection of file system resources. Access restrictions on specific resources and functions are enforced via a fine-grained permission mechanism. Apps are allowed access to resources if they are granted the respective permissions by the user.

This isolation of apps, called sandboxing, is enforced by the kernel, not the Dalvik VM. Java as well as native apps run within a sandbox and are not allowed to access resources from other processes or execute operations that affect other apps.

Apps must declare required permissions for such resources within their app manifest file. These permissions are granted or denied by the user during the installation of the app. The user does not deny or grant permissions during the runtime of the app[5]. Permissions are enforced during the execution of the program when a resource or function is accessed, possibly producing an error if the app was not granted the respective permission. The Android system defines a set of permissions to access system resources such as for reading the GPS location, or for inter-app communication. Additionally, apps may define their own permissions that may be used by other apps.

3 Related Work

Toninelli et al. [9] have investigated current methods of specifying security policies for smartphones. From their assumption, that in mobile computing scenarios users will be required to manage the security on their own, they deduct that the foremost requirement must be to design a simple security model which allows mobile end users to understand their security decisions. Otherwise, this would lead users to define or accept security policies which they do not comply with or also to turn off troublesome security features. Thus, they introduced a semantic-based policy model solution as one step towards a usable security for

---

[5] There is no dynamic permission granting as with the Blackberry system.

smartphones. They assess the efficiency and practicality of their security model
by applying it to typical security related use cases. They have first analyzed
typical mobile use cases and derived critical requirements for the design of a
usable access control model. Their proposed solution relies on the assumption
that understandability of the policy model is a necessary condition for usability
of the access control system.

Their approach relies on a semantic-based policy representation. Such a
semantic-based approach improves the understanding of security policies, they
argue, since users would be more aware of their implications. Although their
work is not directly concerned with apps from the Android Market, their results
can be applied to mobile computing and the smartphone community in particular. However, they also outline that the users tolerance to failure remains a
crucial issue for a usable access control framework.

SMobile has done some research on the Android Market and its permission
system. They have documented specific types of malicious apps and threats. In
their latest paper [10] they have analyzed about 50,000 apps in the Android
Market. They looked for apps which could be considered malicious or suspicious
based on the requested permissions and some other attributes.

Their key findings are that a big number of apps, available from the market,
are requesting permissions that have the potential of being misused to locate mobile devices, obtain arbitrary user-related data and putting the carrier networks
or mobile device at risk. Although the Android Operating System and Android
Market prompt users for permissions before the installation, users are usually
not experienced in making decisions about the permissions they are allowing or
more precisely what permissions an app should have. But most often users do
not take the time or have not the proper knowledge of the security implications.
The most important statement they make is that fundamental security concerns
and increase in malicious apps can be related to poor decisions of the user. Their
work was the most comprehensive security analysis of the Android Market to
date. Their conclusion was that end-users need to make educated decisions regarding the apps they are installing and that third-party security technology
could assist them in making better decisions. This was one of the motivations
for us to make a more in-depth analysis and to provide an open-source framework
for automated permission analysis.

4   Activation Patterns

The analysis of app permissions in the Android Market is based on the *Activation
Patterns* framework that we have developed during the last three years. This
framework has been applied successfully to a wide range of domains such as
event correlation [7], text classification [8] or semantic web analysis [6]. The idea
behind this technique is to transform a raw data vector containing arbitrary
symbolic and real valued features into a pattern, which forms the basis for a
wide range of subsequent analyses. This transformation process is depicted in
Figure 1 which shows several processing layers that:

1. extract features and feature values from instances[6] and store the information as nodes in a semantic network [5],
2. represent relations between these feature values and the strength of these relations (e.g. defined by the number of co-occurences within a data-vector) as weighted links within this network,
3. apply spreading activation techniques [1] for each instance, which stimulate the network and spread the activation of selected nodes according to their links to other regions of the network,
4. and finally extract the activation values for each instance from the network and store them within a vector that we call the *Activation Pattern*.

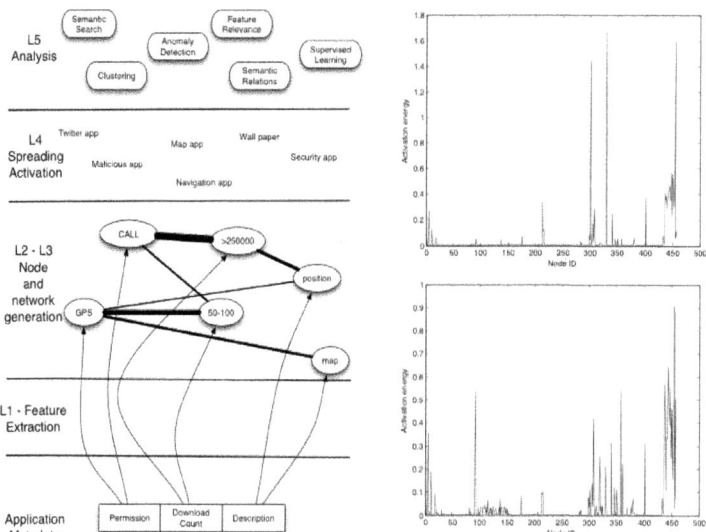

Fig. 1. Left: Layers for the *Activation Pattern* transformation and analysis. **Right:** Examples for two different *Activation Patterns*, the *x*-axis represents the nodes within the network, the *y*-axis represents the activation energy of these nodes after applying the spreading activation process to the activated nodes. These patterns are the basis for all subsequent analyses.

The generated patterns represent the activation values of different regions within the network that are activated due to different input stimuli (e.g. the feature values of an app). The similarity between two patterns can be calculated by distance measures (e.g. the cosine similarity) and expressed as a simple distance value.

---

[6] An instance is a data vector containing all feature values describing the instance. For the Android market, an app instance would be described by various features such as permissions, description terms or download count that have different feature values (e.g. different permissions).

This allows us to apply a wide range of standard machine learning algorithms and thereby cover various analyses with a single model:

Semantic *Search*: The distance between the *Activation Patterns* can be used to implement semantic search algorithms that retrieve semantically related instances. These search queries can also be used to specify certain feature values and find closely related patterns. *Example: Retrieve all description terms and permissions that are semantically related to the GPS permission.*

Feature *Relation*: The semantic network describes arbitrary relations between feature values. By activating one or more nodes (corresponding to feature values) within the semantic network, and spreading their activation via the links to the neighbors, we are able to extract details about the relations between various feature values and the strength of these relations. *Example: Which security permissions are strongly related to the term "GPS"?*

Feature *Relevance*: The relations within the semantic network are created according to the co-occurrence of feature values within the analyzed data set. The strength of these relations are represented by the associated weights within the network. Given a feature value that is represented by a node and the number of emerging/incoming links and their weights, we are able to deduce the importance of the information carried by the node. Nodes that are connected to a large number of other nodes typically do not add information for subsequent analysis processes. *Example: How relevant is the security permission for accessing the GPS when analyzing wall papers?*

Anomaly Detection: The sum of all activation values within a pattern represents the activation energy of the whole pattern, which is a measurement for the response of the network. Anomalies can be detected in two ways: First, if features are combined in a non-typical way, the activation energy is lower than for normal combinations. Second, a large number of inputs (e.g. excessive usage of permissions) causes more activations and thereby higher total activation energies. The first anomaly detection method is more suitable when the number of feature values for each instance is constant. For the market analysis we concentrate on the second method, since the number of features varies from app to app. *Example: Find all wall paper apps that use non-typical permissions.*

Typical Instances: An *Activation Pattern* is characterized by the activation values for the different feature values. By inspecting these values we can determine the strength of the activation and thereby the significance of the feature values. By combining several *Activation Patterns* with simple operations (e.g. mean, variance etc.) we gain knowledge about complete sets of patterns. *Example: What are the typical security permissions of free GPS apps?*

Unsupervised *Clustering*: Due to the transformation into *Activation Patterns* we can directly apply unsupervised clustering algorithms without the need to apply normalization and discretization strategies to the raw feature values. For this work we apply a simple Growing Neural Gas (GNG) algorithm [2] that is extended by the Minimum Description Length (MDL) criterion as used by the

Robust Growing Neural Gas (RGNG) algorithm [4]. This extension enables the algorithm to automatically detect the necessary model complexity, without the requirement to specify the number of clusters in advance[7]. *Example: Cluster all apps with a price larger that 10 Euros according to their description and employed security permissions.*

For a more detailed description of the *Activation Patterns* technique we refer to the appendix of [3].

5 Analyzing the Android Market with *Activation Patterns*

For our analyses we have extracted the metadata of 130.211 apps in December 2010[8]. All subsequent analyses have been conducted according to the following steps:

1. Apply an arbitrary filter to the app database (e.g. apps with certain permission, with a given download count, price or apps that are described with certain keywords).
2. Extract the features and their values from the remaining apps and apply the *Activation Pattern* transformation process.
3. Use the generated patterns for the analyses described in the previous section.

We have found several apps that have either a uncommon combination of permissions or make excessive use of permissions, however we must emphasize that having such permissions does not necessarily mean that the app abuses these permissions. Such abuses cannot be detected by the conducted analyses, but only by inspecting the code of the apps.

Due to space constraints we only highlight some prominent examples that demonstrate the capabilities of the *Activation Patterns* concept and refer the reader to our website where the complete analyses can be downloaded[9].

Q1: Retrieve apps that use the terms "hot" and "girl" in their description and find permission anomalies[10] (**anomaly**): In contrast to the Apple App Store the Android Market allows apps with mature content. Similar to PC software or websites offering such content, we assume that such apps might be infected with trojans, spyware or are built deliberately in order to extract private information from the user.

---

[7] The Activation Patterns concept is not limited to the NG algorithm family – an arbitrary unsupervised algorithm can be applied to the patterns. Obviously, one could also apply supervised algorithms to the patterns for training a classifier, which is not shown in this paper.

[8] The app identifiers where collected from various web sites and the metadata itself was extracted from the Android Market via the android-market-api: http://code.google.com/p/android-market-api/

[9] http://www.carbonblade.at/wordpress/research/android-market/

[10] Hot Girls ALL Without Description.txt.

By applying the ***anomaly*** analysis to the patterns generated for the filtered apps, we are able to gain the following information: A large number of these apps just come with pictures that are displayed within the app or can be set as a wall paper. Some require a connection to the internet in order to grab pictures from web sites. The normal behavior and therefore the typical activation energy within a pattern is defined by this majority of apps. Anomalies deviate from this typical energy and are highlighted by two examples:

The first one is based on a group of apps that describe themselves as apps that "change the picture whenever the user receives an SMS with certain keywords". This description would suggest that the apps are required to have some permissions related to receiving and reading an SMS only, however they make excessive use of permissions related to writing SMSs, reading the contact data, accessing the internet, determining the user's position, using Google auth and many other additional permissions.

The second anomaly refers to an app that includes "hot puzzles and videos". However the app has access to the camera, is allowed to record audio, read and write contact data and has access to the GPS.

Q2: Is there a difference in the typical permissions when comparing free and payed apps with the terms "hot" and "girl" in their description[11] ***(typical instances)?*** In the second example we assume that free apps would be a better target for capturing private information, since they typically have a larger user base. After applying the filters we get the following results for the most active permissions, where the value within the parentheses represents the activation value. A higher value indicates a stronger activation within the network and therefore a higher significance of the corresponding feature value:

- **Payed (644 apps):** *internet* (0.74), *set wallpaper* (0.53), *get tasks* (0.45), *write external storage* (0.45), *get receive completed*[12] (0.42), *access network state* (0.28), *wake lock* (0.23),
- **Free (534 apps):** *internet* (2.95), *set wallpaper* (1.44), *read phone state* (1.19), *access network state* (1.15), *write external storage* (1.06), *access coarse location* (0.95), *access fine location* (0.64), *send sms* (0.46), *read contacts* (0.45), *wake lock* (0.42), *receive boot completed* (0.40), *receive sms* (0.38), *read sms* (0.37), *write sms* (0.37)

These results indicate a gap between free and payed apps with mature content. There might be several reasons for this difference: At first if these apps really include code that capture your private information than it would make more sense to include such code in free apps since they have a larger user base. The second explanation might be found in the light of various ad clients. Developers

---

[11] Hot Girls FREE Without Description.txt and Hot Girls PAYED Without Description.txt.

[12] This permission does not exist in the Android system. It might be a spelling mistake and therefore useless or a self defined permission that is used by multiple apps accessing each other. 88 of the 644 payed apps employ this permission.

often deploy free apps and generate revenue by using ad clients that display advertisements within the app. These ad clients tend to accumulate personal data and often need access to several permissions (see Q9 for more details). The third explanation might be that the developers simply added too many permissions and forgot to remove them. However, this does not seem likely since it would not explain the gap between free and payed apps. To determine what these apps really do, we need to go deeper, decompile these apps and inspect the code and all the calls made to Android APIs.

Q3: Extract popular (more than 5000-10000 downloads) apps with access fine or coarse location permissions and find permissions and terms that occur in the same semantical context as the term "GPS"[13] *(search):* In this example we demonstrate the semantic search capability of the *Activation Patterns* concept. After applying the filter, we get 3204 apps with 5522 terms used for the description. We now execute a semantic search query for the term "GPS" yielding the following results, separated into terms and permissions:

- **Terms:** location, altitude, accuracy, coordinate, track, strength, position, sensor, range, program, technology, compass, satellite, longitude
- **Permissions:** *access fine/coarse location, internet, control location updates, access mock location, wake lock,* followed by permissions for accessing the camera, calling phones, receiving SMS etc.

The related terms are pretty obvious and show that the semantic search queries retrieve significant results. For the permissions it seems that most of the apps need to have access to the internet, prevent the phone from sleeping or dimming the screen (*wake lock*), or simulate a location update (*access mock location*). The last permission is needed especially during development in order to get a position in the emulator and probably was not removed after app deployment.

An example for retrieving semantically related instances would be a query that retrieves apps which do not contain the term "gps" but are semantically related to apps that have the term "gps" in their description (e.g. due to a description that contains "position" or "location"). Other interesting examples are the terms "wife" and "husband" which are closely related to the term "position". This semantic relation is created by apps that are used to spy on someone's wife/husband by tracking her/his location.

Q4: How relevant are the feature values of the apps retrieved in Q3 "GPS" (relevance)?

- **Most Relevant Feature Values:** The most relevant feature values are those that occur rarely. In case of the permissions the most relevant ones are either permissions for special purposes, permissions with a wrong spelling, or self-defined permissions.

---

[13] LOCATION - GPS With Description.txt.

– **Least Relevant Feature Values:** These are values that firstly occur within most of the analyzed apps and secondly that are randomly connected to other feature values. An example is the *internet* permission that is required by a large percentage of apps and is not correlated to other feature values – meaning it co-occurs randomly with those values.

Q5: How is the term "navigation" related to security permissions[14] *(relation)?* In this example we extract the semantic network links to security permission nodes emanating from the node "navigation" and use their weight to determine the strength of these relations: The following permissions are sorted according to the strength of their relation with "navigation": *access coarse/fine location, internet, read contacts, read phone state, write external storage, wake lock, access network state, call phone.* The contact and phone related permissions are typical required in order to control incoming calls/messages from the navigation app. The "navigation" term is also strongly related to the category "travel", high download counts, and GPS relevant terms similar to those of Q4.

Q6: Filter free apps that use the terms "wall" and "paper" in the description and have the permission set wallpaper. Find permission anomalies[15] *(anomaly):* The biggest anomalies are caused by apps that replace the standard launcher of Android. Due to their functionality such apps require a large number of permissions and are therefore considered as anomaly. However, they are followed by several other interesting examples: One of these apps is called *FoxSaver* and describes itself as an app that allows you to browse photos from the website *foxsaver.com* and install them as wall paper. However, the app also has the *receive/read SMS, read contacts, access fine/coarse location* and *call phone* permissions. These additional permissions do not make sense when reading the description of the app.

Q7: Filter apps that have the permissions for reading, receiving and sending SMS messages but do not contain terms related to these permissions in their description (e.g. "sms", "message" etc.)[16] *(anomaly):* The biggest anomalies are caused by apps related to Android development, security and backup. For most of these apps it makes sense to use the permissions, however there are other apps where the description does not match the required permissions: a large number of themes that are just described as "theme" with a certain keyword and several other apps do not state anything about messaging within their description (including games, fitness apps, travel guides etc.).

Q8: Cluster popular apps (more than 250.000 downloads) according to their description and security permissions[17] *(clustering):* After applying the filter, 1079 apps remain that are grouped in the following categories (clusters):

---

[14] LOCATION - NAVIGATION With Description.txt.
[15] Wallpaper Without Description.txt.
[16] SMS PERMS Without Description.txt.
[17] DOWNLOAD COUNT - Only Description.txt and
DOWNLOAD COUNT - Without Description.txt.

- **7 description clusters C1 (55)** ringtone apps; **C2 (64)** apps with German description[18]; **C3 (116)** social network apps; **C4 (339)** tools, widgets, games, browsers; **C5 (123)** translation, reference, books; **C6 (46)** music players, streaming; **C7 (336)** games.
- **8 permission clusters C1 (326)** games with a few permissions (*internet, access network state, read phone state*); **C2 (31)** apps that have access to account data, contacts (e.g. social network related); **C3 (217)** mostly apps with internet access only (reference, games); **C4 (160)** also social network related, but with a bias to location related permissions; **C5 (99)** music and video apps, various permissions, but a strong activation of the *wake lock* permission, which is required to prevent the phone from sleeping; **C6 (59)** mostly ringtones apps with a strong activation of the *read phone state* permission. This permission is required by an app for recognizing that the phone is ringing; **C7 (126)** various apps that require a mixture of different permissions; **C8 (61)** phone and SMS related apps indicated by related permissions.

Especially, the permission cluster results are quite promising, since they allow us to gain knowledge on how permissions are typical used by various application categories and find outliers within a given category.

Q9: Identification and tracking of users: In order read out the unique ID of your smartphone, SIM ID, telephone number or cell ID, the permission *read phone state* is required[19]. In combination with the *internet* permission and possible the location related permissions[20], this enables an app to transmit information which allows user identification and tracking. 31865 of 130211 apps have these two permissions. For the 1079 apps that have more than 250.000 downloads, 362 have these two permissions. This corresponds to the results presented in an analysis of 101 popular apps, which focuses on private data sent to various companies (mostly related to advertisements)[21].

6   Conclusions and Outlook

In this paper we have applied the *Activation Patterns* concept to the Android Market apps. This new technique allows us to extract detailed knowledge about

---

[18] The distance between the German descriptions and the English ones is so large, that the 64 apps are only represented by one cluster. This could be changed by a more complex model.

[19] We refer to `http://developer.android.com/reference/android/telephony/TelephonyManager.html` for a detailed list of extractable information.

[20] The *read phone state* permission grants access to your current cell ID, which could already be used to determine the user's position if an appropriate database containing cell tower locations is available: e.g. `http://www.skyhookwireless.com/`. Therefore, the location can also be determined without the *fine and coarse location* permissions.

[21] `http://online.wsj.com/article/SB10001424052748704694004576020083703574602.html`

the apps and relations between the security permissions, description terms, download counts etc. Since, the Android Market share and therefore the number of apps are growing steadily we argue that the Android platform is an obvious target for malicious activities[22]. For this reason we deem it necessary to get a better understanding of the available apps, their employed security permissions and existing anomalies.

Now, that we have a solid basis for further analysis, we are planning several steps within the next months: The anomaly detection part can be used to screen the market for possible malicious apps, which are then subject to a more detailed analysis. We are currently devising a system that is capable of performing an in depth app analysis.

References

1. Crestani, F.: Application of spreading activation techniques in information retrieval. Artificial Intelligence Review 11(6), 453–482 (1997)
2. Fritzke, B.: A growing neural gas learns topologies. Advances in Neural Information Processing Systems (1), 1211–1216 (2005)
3. Lackner, G., Teufl, P., Weinberger, R.: User Tracking Based on Behavioral Fingerprints. In: Heng, S.-H., Wright, R.N., Goi, B.-M. (eds.) CANS 2010. LNCS, vol. 6467, pp. 76–95. Springer, Heidelberg (2010)
4. Qin, A.K., Suganthan, P.N.: Robust growing neural gas algorithm with application in cluster analysis. Neural Netw. 17(8-9), 1135–1148 (2004)
5. Quillian, M.R.: Semantic Memory, vol. 2, ch.10, pp. 227–270. MIT Press (1968)
6. Teufl, P., Lackner, G.: RDF Data Analysis with Activation Patterns. In: Und Hermann Maurer, K.T. (ed.) Proceedings of the 10th International Conference on Knowledge Management and Knowledge Technologies, iKNOW 2010 Graz Austria, Journal of Computer Science (2010)
7. Teufl, P., Payer, U., Fellner, R.: Event correlation on the basis of activation patterns. In: Proceedings of the 18th Euromicro Conference on Parallel Distributed and NetworkBased Processing, PDP 2010, pp. 631–640 (2010)
8. Teufl, P., Payer, U., Parycek, P.: Automated Analysis of e-Participation Data by Utilizing Associative Networks, Spreading Activation and Unsupervised Learning. In: Macintosh, A., Tambouris, E. (eds.) ePart 2009. LNCS, vol. 5694, pp. 139–150. Springer, Heidelberg (2009)
9. Toninelli, A., Montanari, R., Lassila, O., Khushraj, D.: What's on Users' Minds? Toward a Usable Smart Phone Security Model. IEEE Pervasive Computing 8(2), 32–39 (2009)
10. Vennon, T., Stroop, D.: Android Market: Threat Analysis of the Android Market (2010)

---

[22] ...and also legal apps that identify and track you due to advertisements.

Gesture Authentication with Touch Input for Mobile Devices

Yuan Niu and Hao Chen

University of California at Davis,
Davis, California
{niu,hchen}@cs.ucdavis.edu

Abstract. The convergence of our increasing reliance on mobile devices to access online services and the increasing number of online services bring to light usability and security problems in password entry. We propose using gestures with taps to the screen as an alternative to passwords. We test the recall and forgery of gesture authentication and show, using dynamic time warping, that even simple gestures are repeatable by their creators yet hard to forge by attackers when taps are added.

Keywords: mobile authentication, gestures, android, security.

1 Introduction

There are two prevalent trends in the way we use the Internet today. One is the increasing reliance on mobile devices to access online services, and the other is the ever increasing number of online services. Such services, like online banking, email, social networking, etc, all require its users to create and track credentials. The most common form of credentials are passwords and personal identification numbers (PIN). With so many different services, the number of credentials to remember and manage can be overwhelming, prompting users to fallback on less secure measures such as reusing the same password across multiple sites, creating simpler passwords, or relying on password reset [24].

With smartphones, an additional problem is introduced by the limited screen real estate: *password entry*. Password guidelines suggest that "good" passwords contain a mixture of letters, numbers, and symbols. Memorizing and typing these passwords is already frustrating, but the task becomes even more cumbersome on mobile phones' virtual keyboards where letters, numbers, and symbols may appear on separate screens. A survey of 50 smartphone users revealed that password entry is considered more annoying than other limitations to mobile devices, such as a small screen or poor signal. In addition, 56% of theses users had typed a password incorrectly at least once every ten times [10].

Smartphones using virtual keyboards take input by reading taps to the screen. Often, the user sees a larger or highlighted version of the key being tapped. Password entry fields have also been modified to display the most recently typed letter for a short interval. Soft keyboards often place letters, numbers, and symbols

R. Prasad et al. (Eds.): MOBISEC 2011, LNICST 94, pp. 13–24, 2012.

into separate screens because of space limitations. When they display letters and symbols or numbers concurrently, only a subset of alternate characters are available by holding down the tapped letter for a longer interval to indicate choice of the alternate character. The extra feedback provided helps the user verify her intended keystrokes, but it also helps a nearby third party to observe the user's keystrokes. The extra taps needed to switch between letter, number, and symbol screens also discloses the order of letters, numbers, and symbols within the password.

We propose an alternative to passwords and PINs: *gestures* with touch input. We define gestures as a series of small movements involving motion wrists and forearms while the hand holds the phone. The sheer range of motion and subtleties of force and speed will provide variations between users, even though there some motions, e.g. circles, that may be popular components in gestures. For even more variation, the user can tap the screen with her thumb while she holds the phone to perform a gesture.

2   Gestures

Phishing and human behavior studies demonstrate that humans are creatures of habit [11]. At an even more basic level, muscle memory enables us to learn and perform motor skills quickly and repeatedly without much conscious effort.

Gestures capture a biometric quantity of muscle memory and physical characteristics of the specific user. For the purposes of authentication, we care only about gesture *recall* rather than gestures *identification*.

2.1   Usability Benefits

Gestures provide a faster and more convenient method of authentication compared to a complex password. Studies [2,24] show that passwords that are hard to guess are (not surprisingly) also hard to remember. Gestures, on the other hand, rely on motor skills. Touch typing, brushing your teeth, or signing a signature are examples of fine motor skills into which we put little conscious thought, but are activities we can replicate accurately and precisely.

A few use cases for gesture authentication are described below.

Device Unlock and Second Factor. The most obvious application of gesture authentication is replacing existing phone unlock mechanisms involving pins or gestures drawn on the touchscreen. Gestures can also be used to supplement existing authentication mechanisms where additional proof is needed.

Password Management. Because gestures should be easy to remember, it is possible for users to record one gesture for every unique password. However, remembering which password corresponds to which gesture could get complicated as the number of unique passwords increase. One increasingly common method of dealing with multiple passwords is to rely on a password manager. In this scenario, the gesture will serve as the master password for unlocking all other passwords.

Alternatively, because storing passwords on the phone is risky in case the device is stolen, we could use gestures to access credentials stored remotely[1] or as alternate OpenID credentials. Another advantage of this approach is that it eases the processing burden of running gesture recognition on the phone.

2.2   Security Benefits

A gesture based authentication system would make it more difficult for a shoulder surfer to replay the password, even if he observes the entire gesture. Subtleties like force, speed, flexibility, pressure, and individual anatomical differences would prevent the casual observer from repeating the gesture well enough to authenticate successfully. Furthermore, there are taps to the screen which may be hidden as the user moves the phone. These types of gestures will not require users to look at the screen, so they can gain increased privacy by choosing gestures that can be performed by holding the phone away from the view of onlookers, such as under a table.

Entropy. A text-based password's entropy depends heavily on its length. The entropy per character is $log_2(N)$ bits where N is the size of the pool from which the characters are selected. When all 94 printable ASCII are in the pool, the theoretical entropy per character is 6.55 bits for a random password.

Intuitively, gestures contain more entropy than text passwords because users are not limited to printable characters. Quantitatively, the theoretical entropy provided by our gestures can be measured using the following factors:

It is possible to record with a sample rate between $100hz$ to $120hz$ on some smartphones. The accelerometer in the earliest Android phone can record each axis with $8 - bit$ precision, and this precision will only improve as accelerometers improve. For instance, $13 - bit$ precision recording is already possible on existing accelerometers. Considering orientation gives us 8.49, 8.49, and 7.49 bits of data per sample for the x, y, and z axis, respectively. Using a conservative sampling rate of $60hz$ for both acceleration and orientation, we get 2908.2 bits of entropy per second. If we consider just acceleration, one second of gesture data contains 1440 bits of entropy. Adding touches to the screen further complicate things, as there are any number of taps the user may choose to perform concurrently while performing the gesture.

Shoulder Surfing. Shoulder surfers observing typed passwords on smartphones may have an easier time when the device has a touchscreen, such as an iPhone or Nexus One by taking advantage of visual feedback to the user as they select letters or switch between numbers, symbols, and letters. Gestures are more robust against shoulder surfers, even those with video cameras. It is hard to estimate the force and timing of gestures correctly solely with brute force. Furthermore, the standard recording frame rate for HD video is $29fps$ or $30fps$ in devices like the iPhone, whereas we can sample gesture data at a conservative rate

---

[1] We assume the existence of secure storage for such credentials.

of $60\,fps$. Tapping lightly on the screen with the thumb while moving the phone will further frustrate the attacker, since the attacker would need to identify each tap and its timing.

3   Experimental Design

We conducted two sets of users studies to determine user recall and attacker success rates.

The stages to our experiments are: 1. a) A long term user study with only gestures and b) a follow up attack study and 2. a) an attack study on the two simplest gestures from stage 1 with touches to the screen added and then b) a long term user study on recall. We recruited users from the computer science department at UC Davis because we did not believe that technical expertise would affect gesture choice or performance. We relied on a 3rd party application, Contextlogger [12], which gave us only accelerometer data, for Stage 1, but found that it was insufficient for our needs in Stage 2, at which point we developed a custom Android application.

Stage 1A: Gesture Only.   Each subject was asked to perform the gesture of his choice several times a day, at least 5 iterations at a time, over at least a week. Most subjects participated for at least one month. We had 5 participants who performed 7 gestures. The seven gestures are: 2 are *alphas, the Nike 'swoosh', alpha followed by a circle, first letter of the Thai alphabet, the subject's initials,* and *the subject's signature.*

Stage 1B: Attacks.   We assume an attacker with access to video. The attacker is limited to 5 attempts at gesture replication before the victim's account is locked out. Subjects from Stage 1A were asked if they would allow their gesture to be video recorded. We recorded a total of 6 from 4 participants from a frontal angle and a back angle, which we felt gave the most direct views of the gesture performance.

Participants from Stage 1A were asked to play the role of the attacker with the incentive of a small prize awarded to the most convincing attacker. 6 subjects volunteered. Subjects were allowed as much time as they needed to perform the attacks, and there was no set rules as to how attacks must proceed.

Stage 2A: Adding Taps.   We asked two "victims" from Stage 1B with the simplest gestures, the alpha and the swoosh, to record new videos showing the same gesture they used in Stage 1A, this time with taps added. These gestures were recorded using an Android application that we developed. We recruited 9 subjects to act as attackers. They were asked to view the two victim videos, again with no set rules as to how the attacks must proceed. Attackers were told that in addition to the movement, taps had been added to each gesture. These participants were given the videos to study and when they were ready, asked to record at least ten attempts for each victim gesture.

We chose to perform the attack study before starting a second longterm study to test our hypothesis that adding taps would make very simple gestures harder to imitate.

Part 2B: Extended Study with Taps. We recruited subjects for a long term study, once again, and had 6 users. We asked these subjects to provide at least two weeks of data. The six gestures performed were: the *Nike swoosh* with 5 taps, a *parry-thrust* with 2 taps, *signature* with 4 taps, *two loops* with 3 taps, a *back and forth* motion with 3 taps, and *initials* with 2 taps.

4   Analysis

We present two examples from an earlier feasibility study not discussed in this paper - a simple compound (Fig 1) consisting of an alpha connected to a circle, and a complex gesture (Fig 2) consisting of a Chinese character written in cursive to demonstrate what forgery and recall might look like.

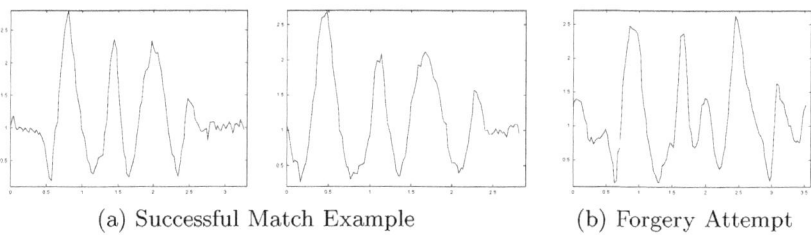

(a) Successful Match Example   (b) Forgery Attempt

Fig. 1. Simple Gesture

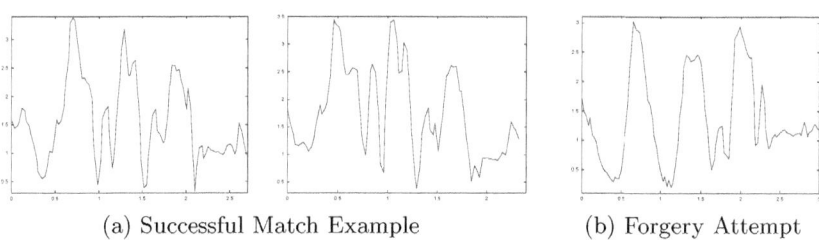

(a) Successful Match Example   (b) Forgery Attempt

Fig. 2. Complex Gesture

As seen in figure 3, multiple trials of the same gesture generally have the same shape, but the timing varies. We used DTW to compute the similarity between a victim's gestures, and the similarity between the victim's gestures and those of his attackers.

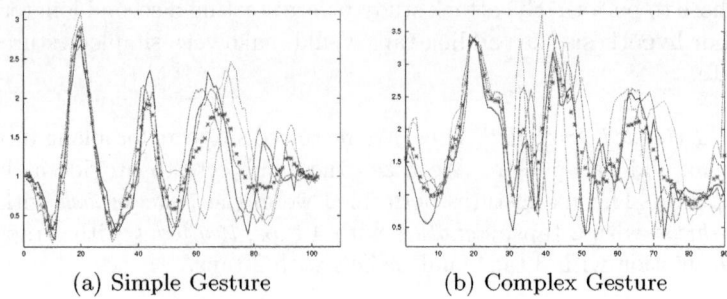

(a) Simple Gesture   (b) Complex Gesture

Fig. 3. Overlays of Gesture Data, adjusted to match for the first feature

4.1 Dynamic Time Warping

Dynamic time warping (DTW) is a dynamic programming sequence alignment algorithm that matches one time series onto a reference time series. The result is a monotonically increasing path and a cumulative matching cost. If two sequences are identical, then the path is a perfect diagonal and the cumulative matching cost is 0. When two sequences differ slightly, the cumulative matching cost is the sum of the distances between the two points matched. DTW with Euclidean distance matching allows us to measure the similarity of two gestures.

4.2 Methodology

In our experiments, we use an existing implementation of DTW [8]. To computer the score for similarity between a gesture and the reference gesture, we do the following:

Let n be the number of repetitions we have for an instance $g \subset G$ of a gesture G, and $r \subset G$ is the set of repeated gestures currently used as reference set. The reference set can be considered the stored "password" and is generated by the user during the initial training period. After this, users only need to perform their gesture once to authenticate. The score, Θ, is then:

$$\Theta = \frac{1}{n} \sum_{i=1}^{n} \frac{1}{n-1} \sum_{j=1}^{n-1} dtw(r_i, g_j) \tag{1}$$

For victim and reference sets, let m be the number of attempts $f \subset F$ we have for a gesture G, and $r \subset G$ is the instance of the victim's gesture currently used as reference. The score, Θ, is then:

$$\Theta = \frac{1}{n} \sum_{i=1}^{n} \frac{1}{m} \sum_{j=1}^{m} dtw(r_i, f_j) \tag{2}$$

We use Equation 1 to generate an average score, which we then used to compute two cutoff values: 1) mean of the repetition scores + standard deviation

2) mean of the repetition scores + half the standard deviation.

The cufoff value determines whether the gesture in question scores low enough (i.e. is similar enough) to the reference gestures.

5 Results

Our extended study showed promising results in that most gestures, except the very simplest, were repeatable by their creators yet difficult to forge by attackers.

5.1 Stage 1: Gesture with No Taps

Gesture Variation. For each user, we gathered at least one week of data (at least 35 repetitions of the same gestures). Our goal in this study was to find a scoring system that is lenient enough to give a legitimate user some leeway in gesture variation and shift over time, and at the same time makes it difficult for an attacker to succeed. We do not address gesture shift over time in our analysis, although we did notice that as time went on, the average DTW scores of later gestures were decreasing and stabilizing.

In these studies, the false negative rate is $1 - rate\ of\ success$ for repetitions, and the false positive rate is the rate of success for attacks. Lower scores indicate greater gesture similarity, because the score is a reflection of the total matching cost between two sequences.

Using Reference Set. Simply using a globally established average DTW value skews the average with initially inconsistent performance. Choosing just one gesture runs into the danger of selecting an outlier that will skew the false negative and false positive rates as well. We randomly sampled 5 files from all gestures to act as our initial, acceptable gesture set and generated a set of $\binom{5}{2}$ DTW scores for each possible pair of gestures. From this set, we tested two possible upper limits to the scores: 1) Average score + Stdev(set of reference scores) and 2) Average score + Stdev(set of reference scores)/2. We calculated average score using Equation 1 where $n = 5$. We adjusted the maximum acceptable score by taking the average of half of the remaining gestures to balance the outliers that were possibly in the initial gesture reference set. We use the same adjusted scores to evaluate the success of attacks. Results are shown in Table 1. This simulates the case where a user has been using gestures to authenticate for some time.

Using adjusted scores, we see that it improves the repetition success rates, but again, skew from outliers is significant enough to improve the attacker's success rates when the gesture is overly simple. A slightly more complicated compound gesture, `alpha+circle`, maintained relatively low false positive and low false negative rates.

Table 1. Repetition Success Rates with Reference Set (1) and Adjustments (2)

	Cutoff 1	Cutoff 2	Attack Cutoff 1	Attack Cutoff 2
alpha1	0.93	0.88	0.57	0.26
alpha2	1	1	0.61	0.26
nike	1	0.98	1.0	0.75
alpha+circle	0.94	0.9	0.06	0.01
initials	0.95	0.93	0.1	0
signature	0.97	0.96	0.15	0.02
Thai letter	0.98	0.93	NA	NA

alpha and circle was successfully performed by just one attacker. Again, the majority of successful attacks on alpha2 came from the subject who performed alpha1.

5.2   Stage 2: Gesture with Taps

Attack Study. To judge success on gesture forgery with taps, we examined the taps data first, as we expected the gesture data to be similar to the attacks described previously. Attackers were often unsure of the number of taps and varied the number as they tried to perform the gestures. At first glance, it seems as though attackers are fairly successful (see Table 2) at guessing the number of taps. However, when we use DTW to score the similarity of timing between attackers and victims, none of the attackers' scores fall into the acceptable range[2].

Table 2. Tapping Count Success Rates for Attacks. The Alpha received 130 attempts, and the Swoosh received 118.

Attacker	Alpha (2 taps)	Swoosh (5 taps)
1	0.21	0.17
2	0.93	0.85
3	0.23	0
4	0	0.70
5	0	1
6	0.57	0.02
7	0.81	0
8	0.08	1
9	0.44	1

[2] Between 0 and (Victim's average score + stdev(victim's score)).

Extended Study. We used the same methodology from stage 1 to evaluate gesture performance. A gesture passes if both the timing and number of taps are correct, and then we evaluate the accelerometer data. We show rates of success in table 3. The results demonstrate that gesture movements are clearly replicable, even with the additional task of memorizing taps, by their respective owner. In some cases, the taps seem to be more difficult to remember than the gesture - for example, the *back-forth* movements were performed with high success rates, but the taps associated with the gesture had the lowest success rates of all gestures. The cutoff rates are adjustable, however, so for services that do not require high security, the more lenient rate could be used for authentication. Conversely, for services that require more security, an even stricter cutoff rate could be enforced.

Table 3. Success Rates of Tasks for Gesturing with Taps

	Touch Count	Touch Timing Cutoff 1	Touch Timing Cutoff 2	Accelerometer Cutoff 1	Accelerometer Cutoff 2
back-forth	0.94	0.87	0.82	0.96	0.89
initials	0.92	1.00	0.97	0.90	0.86
loops	0.97	1.00	0.84	0.92	0.74
parry thrust	0.97	0.98	0.97	1.00	0.79
signature	0.89	0.93	0.79	0.93	0.79
swoosh	0.96	0.94	0.90	0.95	0.88

6 Related Work

Traditionally, authentication factors for computer systems are classified as one of the following: something you know, something you have, or something you are. Biometrics, the measurement of physical characteristics or behavioral traits to identify an individual, are a way to provide evidence for that last factor. Factors such as fingerprints, retinal patterns, or voice patterns have been well studied and evaluated. Jain et al provide a survey of biometric methods [9].

Tapping a rhythm in place of password entry was proposed by Wobbrock [23]. The TapSong user studies showed differences in people's tapping of the same song, and that eavesdropping was difficult because the attacker had no sense of what song was being "played."

Many consumer electronics devices today contain 3-axis accelerometers to measure the positioning and motion of the device. When the device is manipulated by the user, the motions can be a biometric, capturing unique biomechanical traits of the user. Accelerometers have been used to capture gaits [6, 16], arm-sweeps [7], and hand gestures [4, 14, 15]. The accelerometers provide a time-series of measurements in the 3-axes while the motion is performed.

Ravi et al [21] used accelerometer data to recognize 8 activities, such as brushing teeth, running, going up or down stairs, etc using a single accelerometer attached near the pelvic region. They used FFT to extract the energy associated with an activity and used the Weka toolkit [1] to classify the activities. They discussed and tried various classifiers, finally finding that plurality voting yielded the best results.

Existing signature and gesture recognition systems make use of dynamic time warping techniques to score the similarity of inputs [13, 14, 17]. Template adaptation changes the expected sequences used in matching based on age of the template and its similarity to the current input. If the current input is a match, the old template is discarded and the current input becomes the new template. We apply these techniques to address personally unique gesture recognition.

Pylvänäinen [20] used Hidden Markov Models (HMM) to build a recognizer for accelerometer recorded gestures without using feature extractors. However, the gestures described, a "circle or an upward line" are too basic to be used for authentication. They were able to determine that sampling at $30Hz$ was sufficient enough for maximum accuracy.

Similary, Schömer et al [22] worked on gesture recognition and training using the Wiimote. The Wiigee project [19] deals with gesture recognition using a left to right HMM.

Patel et al [18] use gestures to authenticate untrusted public terminals by displaying a pattern that must be replicated on a user's cellphone through gesturing.

Farella et al [5] tested four distinct gestures and found that it is possible to distinguish these gestures in small groups. Chong and Marsden [3] tested the usage of gestures as passwords by creating a limited "alphabet" from which all gesture passwords would be formed. Our system removes the restriction of any alphabet and allows users to choose whatever motions they want as their gesture. Moreover, we are the first, to the best of our knowledge, to combine screen taps with movements to enhance the security of gestures.

Czeskis et al [4] demonstrated that users are able replicate simple gestures accurately in order to activate RFIDs. Again, a key difference in their goals and ours is the reproducibility of gestures: We want only one person to successfully be able reproduce his own gesture.

7   Conclusions and Future Work

We proposed gesture-based authentication on mobile devices. We evaluated several scoring and decision methods using dynamic time warping. We conducted user studies to examine the consistency of repeating one's own gestures over time and the difficulty of emulating others' gestures.

We discovered that the more complicated gestures, unsurprisingly, have low false positive rates and low false negative rates. We also found that we can enhance the security of the simplest gestures by requiring the user to tap the screen during the gesture, because the attackers were unable to observe or replicate correctly the number of taps or the timing of the taps.

Our examination of the gesture with tap data revealed an unexpected benefit: by examining the taps data first, we can sometimes avoid the more computationally expensive task of comparing accelerometer data. Adding taps effectively makes the gesture a two part "password" for attackers, while it remains one integrated motion for the average user.

Gesturing to authenticate can protect users from shoulder surfers and malicious bystanders who may observe the process of password entry. To prevent attackers from emulating gestures, the user should avoid overtly simple gestures, or combine these simple gestures with tapping.

7.1   Future Work

Based on feedback from stage 2 of the user study, one additional step could be implemented before uploading the data: asking the participant whether they feel that gesture could pass the authentication process with a simple "Yes" or "No" popup. We also need to study the effect of feedback to the user in the form of "Pass" or "Fail" when they upload each gesture attempt.

References

1. Weka machine learning project, http://www.cs.waikato.ac.nz/~ml/weka
2. Adams, A., Sasse, M.A.: Users are not the enemy. Commun. ACM 42(12), 40–46 (1999)
3. Chong, M.K., Marsden, G.: Exploring the Use of Discrete Gestures for Authentication. In: Gross, T., Gulliksen, J., Kotzé, P., Oestreicher, L., Palanque, P., Prates, R.O., Winckler, M. (eds.) INTERACT 2009. LNCS, vol. 5727, pp. 205–213. Springer, Heidelberg (2009)
4. Czeskis, A., Koscher, K., Smith, J., Kohno, T.: Rfids and secret handshakes: defending against ghost-and-leech attacks and unauthorized reads with context-aware communications. In: CCS 2008: Proceedings of the 15th ACM Conference on Computer and Communications Security, pp. 479–490. ACM, New York (2008)
5. Farella, E., O'Modhrain, S., Benini, L., Riccó, B.: Gesture Signature for Ambient Intelligence Applications: A Feasibility Study. In: Fishkin, K.P., Schiele, B., Nixon, P., Quigley, A. (eds.) PERVASIVE 2006. LNCS, vol. 3968, pp. 288–304. Springer, Heidelberg (2006)
6. Gafurov, D., Helkala, K., Søndrol, T.: Biometric gait authentication using accelerometer sensor. Journal of Computers 1(7) (2006)
7. Gafurov, D., Snekkkenes, E.: Arm swing as a weak biometric for unobtrusive user authentication. In: International Conference on Intelligent Information Hiding and Multimedia Signal Processing, pp. 1080–1087 (2008)
8. Giorgino, T.: Computing and visualizing dynamic time warping alignments in r: The dtw package. Journal of Statistical Software 31(7), 1–24 (2009)
9. Jain, A., Bolle, R., Pankanti, S. (eds.): Biometrics: Personal Identification in Networked Society. Kluwer Academic Publishers (1999)
10. Jakobsson, M., Shi, E., Golle, P., Chow, R.: Implicit authentication for mobile devices. In: 4th USENIX Workshop on Hot Topics in Security, HotSec 2009 (2009)

11. Karlof, C., Goto, B., Wagner, D.: Conditioned-safe ceremonies and a user study of an application to web authentication. In: Sixteenth Annual Network and Distributed Systems Security Symposium (2009)
12. Kunze, K.: Context logger, http://contextlogger.blogspot.com/
13. Lei, H., Govindaraju, V.: A comparative study on the consistency of features in on-line signature verification. Pattern Recogn. Lett. 26(15), 2483–2489 (2005)
14. Liu, J., Wang, Z., Zhong, L., Wickramasuriya, J., Vasudevan, V.: uWave: Accelerometer-based personalized gesture recognition and its applications. In: IEEE Int. Conf. Pervasive Computing and Communication (PerCom) (March 2009)
15. Liu, J., Zhong, L., Wickramasuriya, J., Vasudevan, V.: User evaluation of lightweight user authentication with a single tri-axis accelerometer. In: Proceedings of the 11th International Conference on Human-Computer Interaction with Mobile Devices and Services, MobileHCI 2009, pp. 15:1–15:10. ACM, New York (2009)
16. Mäntyjärvi, J., Lindholm, M., Vildjiounaite, E., Mäkelä, S., Ailisto, H.: Identifying users of portable devices from gait pattern with accelerometers. In: Proceedings of IEEE Interational Conference on Acoustics, Speech, and Signal Processing, ICASSP 2005 (2005)
17. Nalwa, V.S.: Automatic On-line Signature Verification. In: Chin, R., Pong, T.-C. (eds.) ACCV 1998. LNCS, vol. 1351, pp. 10–15. Springer, Heidelberg (1997)
18. Patel, S., Pierce, J., Abowd, G.: A gesture-based authentication scheme for untrusted public terminals. In: ACM Symposium on User Interface Software and Technology, pp. 157–160. ACM Press (2004)
19. Poppinga, B., Schlömer, T.: wiigee: A Java based gesture recognition library for the wii remote, http://wiigee.sourceforge.net/
20. Pylvänäinen, T.: Accelerometer Based Gesture Recognition Using Continuous HMMs. In: Marques, J.S., Pérez de la Blanca, N., Pina, P. (eds.) IbPRIA 2005, Part I. LNCS, vol. 3522, pp. 639–646. Springer, Heidelberg (2005)
21. Ravi, N., Dandekar, N., Mysore, P., Littman, M.: Activity recognition from accelerometer data. In: American Association for Artificial Intelligence (2005)
22. Schlömer, T., Poppinga, B., Henze, N., Boll, S.: Gesture recognition with a wii controller. In: TEI 2008: Proceedings of the 2nd International Conference on Tangible and Embedded Interaction, pp. 11–14. ACM, New York (2008)
23. Wobbrock, J.O.: Tapsongs: tapping rhythm-based passwords on a single binary sensor. In: Proceedings of the 22nd Annual ACM Symposium on User Interface Software and Technology, UIST 2009, pp. 93–96. ACM, New York (2009)
24. Yan, J., Blackwell, A., Anderson, R., Grant, A.: Password memorability and security: Empirical results. IEEE Security and Privacy 2(5), 25–31 (2004)

A Context-Aware Privacy Policy Language for Controlling Access to Context Information of Mobile Users

Alireza Behrooz[1] and Alisa Devlic[2]

[1] Appear Networks, Kista Science Tower
164 51 Kista, Sweden
alireza.behrooz@appearnetworks.com
[2] Ericsson Research, Färögatan 6
164 80 Stockholm, Sweden
alisa.devlic@ericsson.com

Abstract. This paper introduces a Context-aware Privacy Policy Language (CPPL) that enables mobile users to control who can access their context information, at what detail, and in which situation by specifying their context-aware privacy rules. *Context-aware* privacy rules map a set of privacy rules to one or more user's situations, in which these rules are valid. Each time a user's situation changes, a list of valid rules is updated, leaving only a *subset* of the specified rules to be evaluated by a privacy framework upon arrival of a context query. In the existing *context-dependent* privacy policy languages a user's context is used as an additional condition parameter in a privacy rule, thus *all* the specified privacy rules have to be evaluated when a request to access a user's context arrives. Keeping the number of rules that need to be evaluated small is important because evaluation of a large number of privacy rules can potentially increase the response time to a context query. CPPL also enables rules to be defined based on a user's social relationship with a context requestor, which reduces the number of rules that need to be defined by a user and that consequently need to be evaluated by a privacy mechanism. This paper shows that when compared to the existing *context-dependent* privacy policy languages, this number of rules (that are encoded using CPPL) decreases with an increasing number of user-defined situations and requestors that are represented by a small number of social relationship groups.

Keywords: Context-aware privacy rules, social relationships, mobile users.

1 Introduction

Advances in context-aware technologies are making peoples' lives easier by sensing and collecting information from their surroundings and using this context to assist people in performing their daily tasks. In most scenarios, preserving privacy and integrity of users' personal data is a major issue. People would like to control who can access their context information, at what level of detail, when, and in which

R. Prasad et al. (Eds.): MOBISEC 2011, LNICST 94, pp. 25–39, 2012.
© Institute for Computer Sciences, Social Informatics and Telecommunications Engineering 2012

situations. Therefore, it is important that people can grant restricted access to their context information or deny access to it, depending on their current situation. There is a need to specify a user's *context-aware* privacy preferences, enabling a user to define different sets of privacy rules and when they are applicable. A user's situation is defined by a set of context values that are obtained through some automated means (i.e., via sensors). Note that we distinguish between context values whose access is controlled by a privacy mechanism (i.e., *sensitive context*) and context values which are used to determine a user's situation (i.e., *situational context*).

We tried to express context-aware privacy policies using eXtensible Access Control Markup Language (XACML) [1] in our earlier work [2]. However, for each privacy rule in XACML, a logical combination of conditions can be specified that determines whether the rule should be applied or not. Therefore, we specified a user's situational context in the condition part of the privacy rule along with the authorized requestors' condition and the logical function that should be applied to all condition parts in the rule when the condition is evaluated. Since the logical function can be an AND operation, for a condition to be valid *all* the condition parts have to be evaluated to TRUE. Since at the time when a user situation changes the requestor information is not necessarily available, the requestor condition part cannot be evaluated. Therefore, privacy rules cannot be filtered upon a situational context change. Consequently, upon receiving a context request, *all* the specified privacy rules must be evaluated, *regardless* of the fact that only some of these rules might be valid in the user's current situation. We refer to the described privacy policy languages, in which a user's situational context is used as an additional condition based upon which the decision for granting or denying access to the requested context is made, as *context-dependent* privacy policy languages.

A user's social relation with the requestor has been identified in several studies [3][4] as one of the most important factors influencing a person's willingness to disclose their context information. However, in most of the existing privacy policy languages, social relationship is not used to define privacy preferences. Hence, the ability to define privacy preferences based on a user's social relationships with potential requestors can reduce the number of privacy rules that need to be specified by a user, since potentially a large number of requesters can be represented by a social relation. Consequently, fewer rules need to be evaluated by a privacy mechanism.

In order to address these problems, we propose a *context-aware* privacy policy language (CPPL) based upon the following two design considerations: (1) a user's situations are defined separately from their privacy rules and (2) a context requestor can be specified using its identity or its social relationship to a user. The CPPL design assumes a privacy mechanism that will, when a user's situation changes, select the privacy rules that are valid in this situation. Since a user's situation is defined by a set of values assigned to context parameters, a context change does not necessarily imply a situation change. In fact, we assume that a user's situation will change at least an order of magnitude less frequently than each of its context parameters. Consequently, the selection of a set of privacy rules will be in frequent (as compared to the rate of context changes). When a context request arrives, the privacy mechanism will check (based on a context requestor) only the rules that are valid in the current situation that have been updated at the latest situation change.

We describe architecture of the privacy framework that provides the described context-aware privacy mechanism. Additionally, we create an analytical model to compute the reduction in the number of rules that have to be evaluated by the privacy mechanism when they are specified using CPPL as opposed to when they would be specified using other context-dependent policy languages. We show that this reduction linearly increases with the number of situations and requestors that are represented by a few social relationship groups.

The rest of this paper is organized in 6 sections. Section 2 describes our motivation scenarios, leading to requirements for a *context-aware* privacy policy language. Section 3 reviews the related works according to the identified requirements. Section 4 describes the syntax of the proposed privacy language. Section 5 describes the privacy framework architecture. Section 6 provides the analytical model, while section 7 concludes the paper with plans for future work.

2   Motivation Scenarios and Requirements

This section presents two scenarios that motivate the need for context-aware privacy in the daily lives of average mobile users and derives requirements for a context-aware privacy language from these scenarios.

Bob uses a mobile application that collects his body temperature, heart rate, and blood pressure. A *Healthcare Institution (HCI)* collects *all* these health information of the application users. When a health emergency occurs (i.e., one or more health indicators exceeds a predefined threshold), the application will detect it, then it will send this information along with the Bob's current location to the HCI. HCI will in turn identify the nearest available nurse to Bob and ask her to visit Bob.

Alice uses a mobile application to share her activities with her family and friends on specific occasions. She agrees to allow her husband to see her current activity while she is in Paris, but not otherwise. When Alice is on vacation (this context information can be inferred from her current location and calendar), she wants to inform her friends about the city she is visiting. She is also willing to share her location at the street level with her friends on weekends so they can find each other and go out together.

From these scenarios we identified the requirements that should be considered when selecting an existing or designing a new context-aware privacy policy language:

- *User-defined situation:* A privacy policy language should enable users to define privacy preferences that are valid in specific situations. A situation should be specified by users based on their available context (via a tool with a graphical user interface), using parameters defined in the context model.
- *Rich context model:* A context model should be rich enough to allow users to define any situation, while at the same time it should be customized for use by applications in a particular domain.
- *Periodic-time:* A privacy policy language should enable the definition of privacy rules that are valid during periodic time intervals (e.g., on weekends, on work days from 12:00 to 13:00, etc.).

- *Social relationship:* Users should be able to define their privacy rules based on the social relationship that they have with other people. Maintaining social contacts with a small number of groups and using these groups to specify privacy rules can significantly reduce the number of rules that users need to specify.
- *Fine-grained access:* Users should be able to specify the granularity of context information that they want to disclose to others in a privacy rule.
- *Context-awareness:* Privacy policy rules should be context-aware, thus they should be evaluated when a user's situational context changes (rather than when a request for sensitive information arrives). As a result, only a subset of privacy rules that are valid in the user's current situation will be checked by the privacy mechanism when a context request arrives.
- *Conflict-handling*: There is a potential risk that more than one privacy rule is valid and can be applied in a particular situation. In some cases, these rules can indicate different actions. For example, one rule might grant access to the requested context, while another rule denies it. A privacy policy language must provide a mechanism to handle such conflicts.

3   Related Work

This section reviews the state-of-the-art context-dependent privacy languages according to the requirements identified in the previous section.

3.1   Houdini

Houdini [5] is a context-aware privacy framework that enables users to specify their privacy preferences through web-based forms. Privacy preferences can be defined based on the users' current situation, their social relationship with the requestor or the requestor's identity, and the relation of the requestor's current situation with respect to their own current situation (e.g., if they are located on the same street).

Users can define potential situations through web-based forms. However, situations cannot be defined or modified using the privacy policy language, since the privacy language is decoupled from the context model. Instead, a user's situation is a single variable that is used in the privacy policy rules and whose value must be calculated before evaluating the corresponding rule.

Conflict handling is supported by assigning priorities to rules. If there is a conflict between different rules actions, the rule with the highest priority will be considered.

There is no support in privacy policies for periodic time conditions. Additionally, granularity of disclosed context information cannot be controlled.

3.2   UbiCOSM

The Ubiquitous Context-based Security Middleware (UbiCOSM) [6] represents privacy policies as tuples of one or more contexts that are associated with a set of privacy permissions. A privacy permission determines what kind of operation can or cannot be performed on a particular resource. Privacy permissions are not directly assigned to particular users. Instead, when a user enters a particular context (e.g., a physical location), the associated permission becomes applicable to this user. Permissions have a property that can be assigned either a positive or negative value indicating that access to the requested data is granted or denied, respectively. Additionally, it is not possible to control the granularity of disclosed information.

The UbiCOSM middleware allows mobile users to define situations based on their context and map their privacy permissions to these situations. It updates the set of valid permissions whenever the user's situation changes, which decreases the policy evaluation time when a context request arrives.

There is no explicit support in UbiCOSM for defining a user's situation based on a periodic time interval. Additionally, conflict handling is not supported. A user's social relationship cannot be used (rather than the requestor's identity) to define privacy rules.

3.3   CoPS

The Context Privacy Service (CoPS) [7] enables mobile users to control who can access their context data, when, and at what granularity. CoPS does not enable specification of a user's situations and rules based on a user's context. Instead, CoPS uses an optimistic or pessimistic approach to define a default policy in which all requests are granted or denied, except those that match one of the rules specified by a policy maker. By defining only the rules that specify under which circumstances context should be disclosed or not (depending on the chosen default policy) the number of rules that need to be specified and evaluated is reduced.

Access to context can be granted for the restricted time (e.g., only once, for 2 hours, always allow, or never allow). A context model in CoPS is limited to context variables provided by the middleware. A hierarchical syntax (e.g., "campus.building.room") can be used by a user to control the granularity of the disclosed context information. Context granularity can be specified using a spatial precision (e.g., "Room 123"), temporal restriction, or freshness of the disclosed context information (e.g., to disclose the user's location 15 minutes ago).

CoPS implements definition of groups and access control based on the membership in the specified groups, which decreases the effort of specifying and evaluating the policy rules. Groups can reflect an organization structure or can be defined by a user.

If more than one rule matches the request, conflict handling is performed using the CoPS specificity algorithm that identifies the most specific rule from the matching rules set by comparing their structure fields in the specified order of priority.

3.4   Context Privacy Engine

The Context Privacy Engine (CPE) [8] extends the traditional Access Control List (ACL) mechanism with a set of context constraints that have to be evaluated to validate a particular privacy policy. Context constraints are used to define context conditions that are associated to either the context owner or the context requestor, using XQuery expressions. However, a user's situation defined in the policy is not reusable in other rules. Therefore, if more than one policy should use a particular situation, then this situation definition must be repeated in each of these policies.

A subject and a requestor in CPE policies can be individuals or groups of people. However, a group (e.g., defining a user's social relationships) is expected to be created by an application.

CPE supports conflict handling by considering a policy level, which is an optional field in the policy. A policy at a higher level overrides all policies at lower levels. If there are multiple policies with the same level, the most specific one will be applied.

The CPE policy language does not provide a means to control the granularity of disclosed context information. This policy language is *context-dependant*, thus when a context request arrives, *all* the privacy policies have to be evaluated *regardless* of the user's current situation. Additionally, evaluating *context-dependent* privacy policies requires retrieving context data upon arrival of a context query, which can be time consuming.

3.5   SenTry

The SenTry language [9] is designed as a combination of a user-centric privacy ontology (called SeT Ontology, written in Web Ontology Language (OWL) [10]) and the Semantic Web Rules Language (SWRL) [11] predicates. For each context entity a policy instance is defined which contains the associated privacy rules (defined as SWRL predicates). The SenTry language supports two categories of rules: Positive Authorization Rules (PAR) and Negative Authorization Rules (NAR). NAR rules can only have a "deny" effect, while PAR rules can either allow access to the requested context or transform the context information according to the specified granularity.

Privacy rules are *context-dependent*, thus the context-awareness requirement is not met. Situations can be defined using the SWRL predicates, however SWRL does not support more complex logical combinations of predicates than the conjunction, which makes is difficult to define arbitrarily complex situations.

SenTry provides the "grant override" combination algorithm to handle conflicts among different rules, which is an optimistic algorithm that grants access if at least one rule grants access to the requested context information. Different combination algorithms can be defined to handle conflicts, but it is up to the privacy framework to decide what algorithm should be applied for *all* the policies in the system.

3.6   Summary

Table 1 shows that none of the existing privacy languages fully meets the identified requirements. Most of these languages enable definition of situations and support a rich context model, but none of them enables definition of situations based on periodic time intervals. Languages that satisfy the context-awareness and/or the social relationship requirement enable a small number of privacy rules that have to be evaluated by a privacy mechanism upon a context query arrival.

Table 1. Summary and comparison of context-dependent privacy policy languages

	Houdini	UbiCOSM	CoPS	CPE	SenTry
User-defined situation	+/-	+	-	+/-	+/-
Rich context model	+/-	+/-	-	+	+/-
Periodic-time	-	-	-	-	-
Social relationship	-	-	+/-	+	-
Fine-grained access	-	-	+	-	+
Context-awareness	-	+	-	-	-
Conflict-handling	+	-	+	+	+/-

4   CPPL Model

This section introduces a novel CPPL language that enables mobile users to define their context-aware privacy preferences in a specific granularity based upon the social relationship of a user with a context requestor. By defining different parts of the language, we explain how CPPL meets all the identified requirements.

CPPL specifies *context-aware* privacy rules by mapping a set of *privacy rules* to one or more user *situations*, in which these privacy rules are valid (as shown in Figure 1). A CPPL policy contains one or more context-aware privacy rules.

```
<xs:complexType name="ContextPrivacyRuleType">
  <xs:sequence>
    <xs:element minOccurs="0" ref="cppl:Description" />
    <xs:element type="cppl:Situations" />
    <xs:element type="cppl:RuleSet" />
  </xs:sequence>
  <xs:attribute name="contextPrivacyRuleId" type="xs:ID" />
</xs:complexType>
```

Fig. 1. XML schema representation of a Context Privacy Rule

The *Situations* element contains one or more *Situation* elements each of which is defined by one or more context conditions that must apply to *an* entity (as depicted in Figure 2). To determine if an entity is in a particular situation, all the conditions in the *Conds* element have to evaluate to true. The *Entity* element represents a context owner, which can be an *environment*, a *device*, or a *user*. For example "*<Entity >user|Bob</Entity>*" refers to Bob as a user. The entities in CPPL are expressed using the MUSIC context model [12]. The MUSIC context model enables context parameters to be specified in a hierarchical manner by noting all the parent concepts in the inheritance chain to which the context parameter is assigned to (e.g., "#healthInfo.bloodPressure.systolic"). CPPL uses entities and context parameters to define situations, thus meeting the rich context model requirement.

```
<Situation situationId="Emergency">
<Entity>#user|Bob</Entity><Conds><CondOp op="OR"><Cond>
  <Logical op="OR"><Constraint param="#healthInfo.bloodPressure.systolic"
                              op="NEQ" value="105" delta="15.0"/>
                  <Constraint param="#healthInfo.bloodPressure.distolic"
                              op="NEQ" value="70" delta="10.0"/>
  </Logical></Cond><Cond><Constraint param="#healthInfo.heartRate"
                              op="NEQ" value="75" delta="25.0"/>
  </Cond></CondOp></Conds></Situation>
```

Fig. 2. A Situation element representing Bob's health emergency

Each time a user's situation changes, the list of valid privacy rules are updated; it is this set of rules that will be checked upon receiving a context query. This design makes the CPPL context-aware as was elaborated in section 2.

The *Conds* element contains either a single condition (*Cond*) or a condition operator (*CondOp*) that performs a logical operation on two or more single conditions. Operators provided in the current version of CPPL are logical *AND* and *OR*. The *Cond* element can be defined as a single constraint or a logical combination of constraints. In the example in Figure 2, an "OR" logical combination of the abnormal blood pressure and abnormal heart rate is defined to indicate an emergency situation. The former is a logical combination of two constraints while the latter is a single constraint. Note that two kinds of operators are used in this example. One is applied to constraints or conditions, while another operator is applied to context parameters to define a constraint.

The *Constraint* element (illustrated in Figure 2) is used to specify the set of context parameters that define a condition. It specifies five attributes:

- *entity:* An optional attribute that is used to specify an entity to which the context parameters belong to. If it is not specified, the default entity of parent situation element will be used.
- *param:* It refers to the context parameter that can be assigned a value (e.g., "#location.civilAddress.city").
- *op:* The operator applied to one or two context parameters for constraint verification. Table 2 shows the constraint operators that are supported in CPPL. The definition of these operators is adopted from [13].
- *value:* The value of the context parameter.
- *delta:* This attribute is used for continuous parameters. It shows the acceptable range of context parameters values for a given constraint.

Table 2. Constraint operators

GT	Greater than	NGT	Not greater than	STW	Starts with
LT	Lower than	NLT	Not lower than	ENW	Ends with
EQ	Equals	NEQ	Not equals	NSTW	Not starts with
CONT	Contains	NCONT	Not contains	NENW	Not ends with

The CPPL time constraint element is used to specify any (periodic) time constraint. It contains either a *DateRange* element or an *Interval* element. The former defines a time range that *begins* and *ends* at the specified date and time. The latter specifies an interval that has the following attributes:

- *daysOfWeek*: denotes week days in the form of numbers or words, representing one or more days or a range of days (e.g., notations "mon,wed,fri" or "1-3" can both be used to represent Monday, Wednesday, and Friday).
- *months*: denotes months of the year by their names or numbers. Months can enumerated or represented by a range (e.g., "2,4,7" or "may-aug").
- *daysOfMonth*: numbers between 1 and 31 that indicate days in a month.

The *Interval* element contains an optional "*TimeRange*" element for specifying time periods with "*startTime*" and "*endTime*" attributes. An example of using the *TimeConstraint* element to represent working hours is shown in Figure 3.

```
<TimeConstraint ><Interval daysOfWeek="MON-FRI">
    <TimeRange startTime="08:00:00" endTime="16:00:00" />
</Interval></TimeConstraint>
```

Fig. 3. An example of the use of the *TimeConstraint* element

Figure 4 shows the definition of *Ruleset* element, a collection of *Rule* elements that are mapped to one or more situations. The *Rule* element (see an example in Figure 7) describes who (*Identity* element) can access what kind of context information (*ContextParams* element). The effect of a privacy rule is to *permit* or *deny* access to the requested context information. When there is more than one rule in a rule set, a combination algorithm should be used to evaluate the final effect and resolve potential conflicts of the RuleSet. This algorithm uses a "denyOverrides" or "permitOverrides" policy to *deny* or *permit* access to the requested context if at least one rule from the set evaluates to *deny* or *permit*, respectively. This algorithm can be used to determine the final effect of the multiple rule sets in a user's privacy policy.

```
<xs:complexType name="RuleSet">
   <xs:attribute name="combinationAlg" type="xs:string" use="optional" />
   <xs:sequence><xs:element type="cppl:RuleType" /></xs:sequence>
</xs:complexType>
<xs:complexType name="RuleType"><xs:sequence>
   <xs:element type="cppl:IdentityType" /> <xs:element type="cppl:ContextParamsType" />
   </xs:sequence>
   <xs:attribute name="effect" type="cppl:EffectType" />
</xs:complexType>
```

Fig. 4. XML schema representation of Ruleset element

The *ContextParams* element specifies sensitive context parameters assigned to an entity. Using an *AnyContextParam* element within the *ContextParams* element indicates that all context parameters can be accessed by all potential requestors.

Privacy rules can be defined based on a user's social relationship with a context requestor, enabling a user to specify a rule for a *class* of requestors instead for each requestor individually. The *Identity* element enables different ways of representing a context requestor in a privacy rule, using the following elements:

- *One:* an individual represented by an id.
  - e.g. <one id="sip:admin@HCI.com"/>
- *Many:* a group of users in the same administrative domain.
  - e.g. <Many domain="HCI.org"/>
- *Relation:* a group of people having a specific relation to the context owner.
  - e.g. <Relation relation="spouseOf"/>
- *AnyIdentity:* is used for rules that should be applied to all the requestors.

Figure 5 shows an example of *RuleSet* element that permits Alice's friend to access her current location at the city level.

```
<Rule effect="Permit">
    <Identity><Relation relation="friendOf"/></Identity>
    <ContextParams><ContextParam>
    <Entity>#user|Alice</Entity>
    <Param>#location.civilAddress.city</Param>
    </ContextParam></ContextParams>
</Rule>
```

Fig. 5. Privacy rule allowing Alice's friends to access her location

When a group of people are selected using the *Many* or *Relation* elements, it is possible to exclude one ore more individuals from the selected group using the *Except* element (as depicted in Figure 6).

```
<Identity><Many domain="HCI.org">
    <Except id="sip:Alice@example.com"/>
</Many></Identity>
```

Fig. 6. Using Except element to exclude an individual from a group

5   Privacy Policy Management System Architecture

A *Privacy Policy Management System (PPMS)* provides a context-aware privacy mechanism supporting the CPPL language that enables a user to control access to his/her sensitive context depending on this user's current situation. We refer to the user controlling access to his/her sensitive context information as the *context owner* and to the user requesting access to the context owner's sensitive context as the *context requestor*. Note that we assume that a context owner and a context employ different devices to control and request access to a particular context, respectively.

As illustrated in Figure 7, the PPMS executes at the context owner's device and consists of four components: *Policy Administration Point (PAP), Policy Refinement Point (PRP), Policy Enforcement Point (PEP), and Policy Decision Point (PDP)*. A context owner specifies his/her privacy preferences (step 1). These preferences are transformed by the PAP into the CPPL policies and stored into the *CPPL Repository* (step 2). The PRP reads the privacy rules from the policies stored in the CPPL repository (step 3), extracts the *situational context parameters* that are used in conditions of these rules, and sends the request to the *Context Provider* to retrieve these context parameters (step 4). The *Context Provider* obtains the requested context from the sensors that can provide this information and fires the context changed event containing the sensed context data. This event is received by all the components that have previously requested this information (steps 5 and 11).

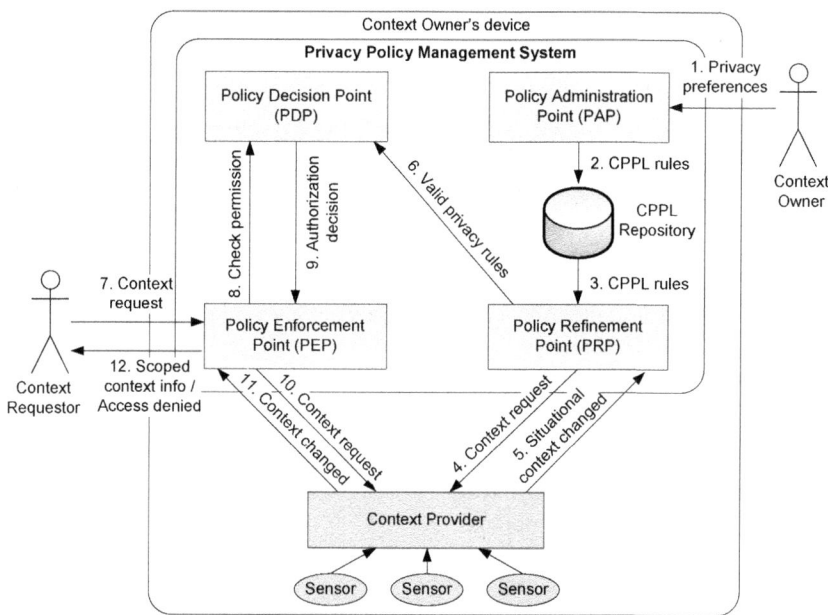

Fig. 7. The overall architecture of privacy policy management system

The PEP controls access to the context owner's sensitive context. After receiving a context request from the context requestor (step 7), the PEP asks the PDP to check if this requestor is permitted to access the requested context (step 8). The PDP checks the list of valid rules in the owner's current situation in order to determine is the requestor is permitted to access the requested *sensitive context*. It returns the authorization decision to the PEP, which is either to *permit* or to *deny* access to the requested context (step 9). If the request is permitted, the PEP will forward the context request to the context provider (step 10). Otherwise, the context request will be rejected. After obtaining the context information from sensors, the PEP will receive the updated context (step 11), format this context in the granted scope (if the access to it has been permitted), and send this scoped value to the requestor (step 12).

If the updated context sent in the changed context event is the situational context, it will be received by the PRP (step 5). The PRP will determine if this situational context update has caused the change of a user's situation and if so it will update the list of valid privacy rules in this new situation & send this list to the PDP (step 6). This process is called *policy refinement* and must be performed before determining if the context requestors are authorized to access the requested context. However, if the changed context is not a situational context, the policy refinement is not necessary because the valid rules are already up to date.

The PPMS has been implemented in Java and integrated as a component in the MUSIC context middleware [14]. We plan, as part of the future work, to evaluate the performance of the proposed privacy mechanism in terms of the latency that this mechanism adds to the context response time.

6   Analytical Model

As earlier elaborated, a small number of privacy rules encoded using a context-aware privacy policy language should be evaluated by a privacy mechanism upon arrival of a context query. To fulfill this goal, CPPL introduces two approaches: (1) filtering the context-aware privacy rules based on the context owner's current situation and (2) representing context requestors using their social relationship with the context owner.

This section describes an analytical model that is used to compute and compare the number of privacy rules that need to be evaluated by a privacy mechanism when they are specified using a context-dependent privacy policy language vs. when they are specified using the CPPL.

In order to simplify the model we made the following three assumptions:

- Each privacy rule specifies access of a single requestor (i.e., an individual or a group of people) to a single context parameter.
- For any requestor and the owner's sensitive context parameter, a separate privacy rule has to be defined.
- The number of the sensitive context parameters whose access is controlled is the same in each situation.

In *context-dependent* privacy policy languages, the number of privacy rules that need to be evaluated by a privacy mechanism (N_{CB}) is equal to the number of requestors used in the privacy rules (R) multiplied by the sum of the number of sensitive context parameters (C_{Si}) in all the user's defined situations (S):

$$N_{CB} = \sum_{i=1}^{S} R * C_{S_i} = R * \sum_{i=1}^{S} C_{S_i} \qquad (1)$$

Since the number of sensitive context parameters in different situations is the same (i.e., $\forall i, j \in [1, S]$, $S \in N, C_{S_i} = C_{Sj} = C_S$) (1) becomes:

$$N_{CB} = R * S * C_S \qquad (2)$$

In CPPL, the number of privacy rules that need to be evaluated by the privacy mechanism (N_{CPPL}) is equal to the number of valid privacy rules in a user's current situation. Since the privacy rules in CPPL can be specified for both the individual requestors and the groups, N_{CPPL} can be calculated as the sum of the number of groups (G) and number of individual requestors that are identified by their identity (R_{IND}) multiplied with C_S:

$$N_{CPPL} = (G + R_{IND}) * C_S \qquad (3)$$

In order to compare the number of rules that need to be evaluated by a privacy mechanism when they are encoded using a context-dependent privacy language vs. when they are encoded using CPPL, we compute the ratio of N_{CB} and N_{CPPL}:

$$\frac{N_{CB}}{N_{CPPL}} = \frac{R * S}{G + R_{IND}} \qquad (4)$$

The result of (4) can be interpreted as the reduction in the number of the rules achieved by CPPL design of context-aware privacy rules. Since each potential requestor either belongs to a group or is considered as an individual requestor, and assuming that there are no empty groups, the sum of groups and individual requestors is *less than or equal to* the number of all requestors. Furthermore, there is always at least one situation even if a user has not defined any situation, to which the privacy rules are mapped (i.e., an *"Always"* situation). Thus, it can be concluded that N_{CPPL} will always be *lower than or equal to* N_{CB}. Additionally, the more user-defined situations and the larger the number of requestors that are represented by social relationship groups, the more effective CPPL becomes compared to the existing context-dependent privacy policy languages. If all the rules are defined for individual requestors and there is only one situation, the number of rules in CPPL will be equal to the number of rules in context-dependent privacy languages.

7   Conclusion

This paper introduces a context-aware privacy policy language (CPPL), which can be used to represent context-aware privacy preferences of mobile users in order to control access to their context information. A user's context is used in context-aware privacy rules to specify which of these rules are valid in particular situation(s). CPPL enables a user to define situations using a set of context parameters that are defined in a context model. When a user's current situation changes, a list of valid rules is updated by a privacy mechanism (that implements support for CPPL), thus leaving only relevant rules to be evaluated upon arrival of a context query.

CPPL enables a user to specify privacy rules based on a social relationship with a context requestor, thus reducing the number of rules that need to be specified by the user and that consequently need to be evaluated by the privacy mechanism.

In the existing context-dependant privacy policy languages a user's context is used as an additional condition in a rule, along with a requestor's identity. In order to process a context request, the privacy mechanisms supporting these languages need to process *all* the specified rules and to retrieve all context values that are used to define privacy preferences, which can in the case of evaluation of a large number of rules, significantly increase the privacy policy evaluation time.

We provide an analytical model that calculates a reduction in the number of rules that need to be evaluated by a privacy mechanism when they are encoded using CPPL (vs. when they are compared to context-dependent privacy languages), showing that effectiveness of CPPL increases with an increasing number of defined situations and requestors that are represented by a small number of social relationship groups.

As part of the future work, we plan to perform a performance evaluation of the proposed privacy framework in terms of the latency it adds to the context response time. Moreover, a usability study will be done to make sure that our approach can be easily employed by average mobile users.

References

[1] Moses, T.: eXtensible Access Control Markup Language (XACML) Version 2.0. Technical report, OASIS (February 2005)

[2] Devlic, A., et al.: Context inference of users' social relationships and distributed policy management. In: Proc. of the 7th IEEE International Conference on Pervasive Computing and Communication (PerCom 2009), 6th Workshop on Context Modeling and Reasoning (CoMoRea 2009), Galveston, Texas, USA, pp. 755–762 (March 2009)

[3] Consolvo, S., et al.: Location Disclosure to Social Relations: Why, When, and What People Want to Share. In: 11th International Conference on Human-Computer Interaction (CHI 2005), pp. 81–90. ACM Press, Las Vegas (2005)

[4] Olson, J.S., et al.: Preferences for Privacy Sharing: Results & Directions CREW Technical Report (2004)

[5] Hull, R., et al.: Enabling context aware and privacy-conscious user data sharing. In: 5th IEEE International Conference on Mobile Data Management (MDM 2004), Berkley, CA, USA, pp. 187–198 (January 2004)

[6] Corradi, A., Montanari, R., Tibaldi, D.: Context-based Access Control Management in Ubiquitous Environments. In: Third IEEE International Symposium on Network Computing and Applications (NCA 2004), Cambridge, MA, USA, pp. 253–260 (August 2004)

[7] Sacramento, V., Endler, M., Nascimento, F.N.: A Privacy Service for Context-aware Mobile Computing. In: First International Conference on Security and Privacy for Emerging Areas in Communications Networks (SecureComm 2005), Athens, Greece, pp. 182–193 (September 2005)

[8] Blount, M., Davis, J., et al.: Privacy Engine for Context-Aware Enterprise Application Services. In: IEEE/IFIP International Conference on Embedded and Ubiquitous Computing, Shanghai, China, vol. 2, pp. 94–100 (December 2008)

[9] Czajkowski, K., Fitzgerald, S., Foster, I., Kesselman, C.: Grid Information Services for Distributed Resource Sharing. In: 10th IEEE International Symposium on High Performance Distributed Computing, San Francisco, pp. 181–184 (2001)

[10] McGuinness, D.L., Harmelen, F.: OWL web ontology language overview. W3C submission, W3C Recommendation (2003), http://www.w3.org/TR/owl-features/

[11] Horrocks, I., et al.: SWRL: A Semantic Web Rule Language Combining OWL and RuleML. W3C submission, http://www.w3.org/Submission/SWRL/

[12] Reichle, R., Wagner, M., Khan, M.U., Geihs, K., Lorenzo, J., Valla, M., Fra, C., Paspallis, N., Papadopoulos, G.A.: A Comprehensive Context Modeling Framework for Pervasive Computing Systems. In: Meier, R., Terzis, S. (eds.) DAIS 2008. LNCS, vol. 5053, pp. 281–295. Springer, Heidelberg (2008)

[13] Reichle, R., et al.: A Context Query Language for Pervasive Computing Environments. In: Sixth Annual IEEE International Conference on Pervasive Computing and Communications (PerCom 2008), Hong Kong (March 2008)

[14] IST project MUSIC, Self-Adapting Applications for Mobile Users in Ubiquitous Computing Environment project, http://www.ist-music.eu

Android Security Permissions –
Can We Trust Them?

Clemens Orthacker, Peter Teufl, Stefan Kraxberger, Günther Lackner,
Michael Gissing, Alexander Marsalek,
Johannes Leibetseder, and Oliver Prevenhueber

University of Technology Graz,
Institute for Applied Information Processing and Communications, Graz, Austria
{clemens.orthacker,peter.teufl,
stefan.kraxberger,guenther.lackner}@iaik.tugraz.at

Abstract. The popularity of the Android System in combination with
the lax market approval process may attract the injection of malicious
applications (apps) into the market. Android features a permission system allowing a user to review the permissions an app requests and grant
or deny access to resources prior to installation. This system conveys a
level of trust due to the fact that an app only has access to resources
granted by the stated permissions. Thereby, not only the meaning of
single permissions, but especially their combination plays an important
role for understanding the possible implications. In this paper we present
a method that circumvents the permission system by spreading permissions over two or more apps that communicate with each other via arbitrary communication channels. We discuss relevant details of the Android
system, describe the permission spreading process, possible implications
and countermeasures. Furthermore, we present three apps that demonstrate the problem and a possible detection method.

Keywords: Android Market, Security Permissions, Android Malware,
Android Services, Backdoors, Permission Context, Side Channels.

1 Introduction

The opening of mobile device platforms to third party developers was probably
the most significant move of the IT industry in the last years. The availability of
a multitude of applications (apps) has boosted user acceptance and usefulness of
mobile devices like smartphones and tablet computers. Regardless whether for
business or private use, these devices and their apps have the potential to facilitate and enrich the user's everyday life. Mobile platform vendors recognized the
importance of opening their systems to third party developers in order to attract
a wide range of apps. However, the large number of third party developers make
it very hard to provide uniform quality standards for the repository providers.

While, i.e., Apple enforces tight policies on software distributed via their App-
Store for iOS, regarding security and content, Google emphasizes a more *open*

R. Prasad et al. (Eds.): MOBISEC 2011, LNICST 94, pp. 40–51, 2012.

philosophy, providing many liberties to Android developers, distributing their products via the Android Market. Apps submitted to the Android Market are rudimentarily checked but the process is not as strict as it is for the AppStore. Google seems to pursuit a *delete afterwards* strategy if apps have been found, which are malicious or are of low quality. Android implements a kill switch to remotely remove apps installed on customer devices[1].

Google introduced a permission system for their Android platform, allowing developers to define the necessary resources and permissions for their products. The customer can decide during the installation whether she wants to grant or deny access to these requested resources such as the address book, the GPS subsystem or the phone functionalities. Although, this process is challenging to the standard user, at least an expert will have the ability to draw conclusions about the theoretical capabilities of an app based on its permissions.

Thereby, the security of the permission system is mainly based on two different aspects: the meaning of the permission itself and even more important, the meaning of combined permissions. For example, when the Internet permission is combined with the read contacts permission, a possible malicious app could transfer your private contact data to the Internet. This implication and the functionality is lost when both permissions are not used in the same application. Therefore, a large part of the security of the permission system and the trust in it is based on the assumption that an app only gains access to the resources that are declared via permissions.

In this work we give a detailed description of the possible communication paths between applications. We discuss the issue of what we call *spreading of permissions* which exploits interprocess communication to allow a transfer[2] of security permissions to apps which did not request them at installation time. We substantiate this threat by presenting three prototype apps that highlight the permission spreading problem and demonstrate the detection of a Service based communication path[3].

2   Related Work

In the article *Understanding Android Security* [1] Enck et al. took a look at the Android application framework and the associated features of the security system. One pitfall of Android, as the authors describe, is that it does not provide information flow guarantees. Also the possibilities of defining access policies in the source code introduces problems because it clouds the app security since the manifest file does not provide a comprehensive view of the application's security anymore.

---

[1] http://www.engadget.com/2010/06/25/google-flexes-biceps-flicks-android-remote-kill-switch-for-the/

[2] This is not an actual transfer of the permission, but has the same effect.

[3] These apps can be downloaded from:
http://www.carbonblade.at/wordpress/research/android-market/

SMobile has done some research on the Android Market and its permission system. They have documented specific types of malicious apps and threats. In their latest paper [6] they have analyzed about 50,000 apps in the Android Market. They looked for apps which could be considered malicious or suspicious based on the requested permissions and some other attributes.

Their key findings are that a big number of apps, available from the market, are requesting permissions that have the potential of being misused to locate mobile devices, obtain arbitrary user-related data and putting the carrier networks or mobile device at risk. Although the Android OS and Android Market prompt users for permissions before the installation, users are usually not ready to make decisions about the permissions they are granting. The most important statement they make is that fundamental security concerns as well as increase in malicious apps can be related to poor decisions of the user. Toninelli et al. came to the same conclusion in [5].

Nauman et al. [2] investigated the Android permission system with special focus on introducing more fine-grained permission assignment mechanisms. The authors argue that the current permission system is too static since it does not take into account runtime constraints and that the accept all or none strategy is not adequate. Thus, they propose Apex, an extension to the Android policy enforcement framework which allows users to grant permissions more selectively as well as to impose constraints on the usage of resources at runtime. Their extension provides some additional security features which allows a more fine-grained security policy for Android. A similar approach is outlined in the work of Ongtang et al. [3].

Similar to [1], Shabtai et al. [4] performed a security assessment of the Android framework in the light of emerging threats to smartphones. They made a qualitative risk analysis, identified and prioritized the threats to which and Android device might be exposed. In addition, they outlined the five most important threat categories which should be countered by employing proper security solutions. They provide a listing with adequate countermeasures and existing solutions for the specific threat categories. One of the main propositions is that the permission system should be hardened to protect the platform better from misuse of granted permissions.

3   Interprocess Communication

Android apps may activate components of any other app if the other app allows for it.

The four different types of components providing entry points for other apps are Activities[4], Services, Broadcast Receivers and Content Providers.

If a certain component is requested, the Android system checks whether the corresponding app process is running and the component is instantiated. If either of both is not available, it is created by the system. Thus, if a requesting app is

---

[4] For all subsequent Android specific terms we refer to the Android developer documentation located at http://developer.android.com/index.html

allowed to access a background Service provided by another application, it can access it at any time, without the Service having to be started by the user.

3.1   RPC Communication

The Android system provides means for interprocess communication (IPC) via remote procedure calls (RPC). Since different processes are not allowed to access each other's address space, methods on remote objects are called on proxy objects provided by the system. These proxy objects decompose (marshal) the passed arguments and hand them over to the remote process. The method call is then executed within the remote app component and the result marshaled and returned back to the calling process. The app programmer merely defines and implements the interface. The entire RPC functionality is generated by the system based on the defined interface and transparent to the application.

Interfaces for interprocess communication are defined using the Android interface definition language (AIDL). The resulting Java interface contains two inner classes, for the local and remote part, respectively.

Typically, the remote part is implemented within an app component called *Service*, allowing clients to *bind* to it in order to receive the proxy object for communication with the remote part. The Service returns the Stub class in its `onBind()` method called by the system upon a connection request from the client. The client, on the other hand provides a `ServiceConnection` callback object along with the bind request in order to receive the proxy object for interprocess communication with the Stub.

Specific messages, called *Intents*, identifying the targeted Service, represent bind requests. A client's Intent is passed to the Service's `onBind()`, so that the Service can decide whether or not to accept the connection. Upon a successful connection establishment, the system passes the proxy object corresponding to the Stub returned from `onBind()` to the client's `ServiceConnection` callback.

Services may declare required *permissions*[5] that are enforced during the binding. Applications binding to the Service have to declare the use of these permissions correspondingly.

3.2   Communication via Intents

Intents are logical descriptions of operations to perform and are used to activate Activities, Broadcast Receivers and Services. Since these components may be part of different applications, Intents are designed to cross process boundaries and may transmit information between applications. Note that Intents are mainly intended to identify a component, optionally adding a limited amount of additional information to more precisely specify the targeted operation.

Neither Activities, nor Broadcast Receivers or (unbound) Services provide persistent RPC connections for interprocess communication. Still, they may all be activated by Intents which will be passed to their respective activation methods

---

[5] See Section 4.

`startActivity()`, `startService()`, `sendBroadcast()` and others. Via their `extras` attribute, Intents therefore provide a simple means for transmitting a limited amount of data to another process' component. Additionally, Activity components provide a way to return a result back to their caller. If launched via `startActivityForResult()` the calling process may receive a result Intent via its `onResult()` method, thus allowing for simple two-way communication.

3.3   Alternative Communication Paths

Content Providers are intended to share data between applications. They are uniquely identified by URIs and can be accessed via `ContentResolver` objects provided by the system. Note that Content Providers are not activated by Intents. Reading and writing to Content Providers allows for two-way interprocess communication if the participating processes have the required permissions.

Apart from the described methods, apps may also communicate by exposing data via the filesystem and setting global (world) read/write permissions on the files. Alternatively, apps may share resources if they request the same UID. In that case they are treated as being the same app with the same file system permissions. UID sharing is possible for apps signed by the same developer.

We consider these communication approaches as *side-channel* communication.

3.4   Information Flow Overview

Based on the IPC methods described above, there are four possible communication paths between two apps A and B (see Figure 1). The S block depicts any of the aforementioned interprocess interfaces that provides or receives information. The arrows indicate the information flow between an app and the remote component.

- **One-Way:** An app A can either transfer/receive information to/from an app B, by using a one-way communication method. This could also be described as pushing or pulling information to/from an application.
- **Real Two-Way:** App A exchanges information with app B, by using two-way communication. Thereby both apps can receive and transmit information from A to B and vice versa. This involves an RPC interface as provided by Services that a client can bind to.
- **Pseudo Two-Way:** In this variant, two one-way communication channels are combined to create a two-way communication channel. In this example, app A transmits information to app B via interface S2 provided by B. In addition, app B pushes information to A by calling its S1 interface. For this example two push channels were used, however an arbitrary combination of push and pull channels could be used. On an abstract information flow level, this method is equal to the real two-way communication method (regardless of the employed push/pull combination).

Especially the pseudo two-way method and one-way push method can be used to transmit information over side channels (e.g. communication by reading and writing to logging facilities). The available communication paths influence how potential malicious apps can avoid detection and where they locate the actual malicious code.

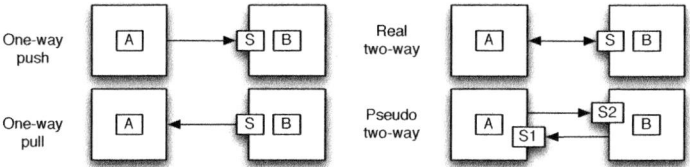

Fig. 1. Two apps A and B can exchange information via one-way or two-way communication

4   Android Permission Mechanism

Application isolation, distinct UIDs for all apps and permissions are the three building blocks of Android's security architecture. Isolation of apps from each other as well as from the system is assured by executing every app within its own Linux process. Further, every app runs with a distinct user- and group ID (UID), assigned at app installation. This allows for protection of memory and file system resources. Communication and resource sharing are subject to access restrictions enforced via a fine-grained permission mechanism. Applications are allowed access to resources if they are granted the respective permissions by the user.

The isolation of applications, called *sandboxing*, is enforced by the kernel, not the Dalvik VM. Java as well as native apps run within a sandbox and are not allowed to access resources from other processes or execute operations that affect other apps.

Applications must declare required permissions for such resources within their manifest file. These permissions are granted or denied by the user during the installation of the application. The user does not deny or grant permissions during the runtime of the application[6]. Permissions are enforced during the execution of the program when a resource or function is accessed, possibly producing an error if the app was not granted the respective permission. The Android system defines a set of permissions to access system resources such as for reading the GPS location, or for inter-application/process communication. Additionally, apps may define their own permissions that may be used by other apps.

There is no central point for permission enforcement, it is scattered over many parts of the Android system. At the highest level, if permissions are declared in an application's manifest file for a component, they are enforced at access points to that component. These are calls to `startActivity()` or `bindService()` for

---

[6] There is no dynamic permission granting as with the Blackberry system.

activities or services, respectively, that would cause security exceptions to be thrown if the caller is not granted the required permission.

Permissions may control the delivery of broadcast messages by restricting who may send broadcasts to a receiver or which receivers may get the broadcast. In the first case a permission for the protected receiver is declared in the manifest file (or when registering the receiver programmatically, respectively). It gets enforced *after* a sender's call to sendBroadcast() and will not cause an exception to be thrown. Rather, the message will simply not be delivered to that receiver if the sender does not have the required permission. A sender, on the other hand, may declare a permission within the sendBroadcast() call, which will also be enforced without the sender noticing.

Permissions for granting read or write access to Content Providers are declared within the manifest file. Apart from that, content providers allow for a finer grained access control mechanism, via URI permissions. They control access to subsets of the content provider's data, allowing a content provider's direct clients to temporarily pass on specific data elements (identified by a URI) to other applications. A dedicated flag on an Intent[7] indicates that the recipient of the Intent is granted permission to the URI in the Intent's data field, identifying a specific resource, such as a single address book entry. The granted URI permission is finally enforced once the recipient of the Intent queries the content provider holding the URI by calling on a ContentResolver object[8]. Content Providers declare support for URI permissions in the manifest file. Enforcement of URI permissions results in security exceptions being thrown if the caller does not have the required permissions.

Applications may at any time query their context whether a calling PID or package (name) is granted a permission. This allows for custom-tailored permission enforcement for specific app requirements. Certain system permissions are mapped to Linux groups. On app installation, the application's UID is added to the respective group (GID). Permission enforcement involves GID checks on the underlying OS level. The permission to GID mapping is declared within the system's platform.xml file. Another specificity are *protected broadcasts*, which only the system may initiate.

5   Permission Spreading

As the user grants permissions on installation of an application, it is crucial to consider all permissions an app requests *in context*. The combination of specific permissions may indicate security flaws to the user. Within this work, we consider the user to be capable of critically analyzing an application's declared permissions and understand the implications of granting permissions.

Therefore, the permission system and the decision of the user is based on the assumption that an app can only use the functionality for which the appropriate permissions are available. We argue, that this security function can

---

[7] Intents are the entities used to activate app components, cf. Section 3.2.

[8] Content Providers are not activated via Intents.

be circumvented by spreading security permissions over two or more apps that use interprocess communication. Thereby the apps are able to gain additional functionality for which they do not have the corresponding permissions.

5.1 Demonstration

In this section we describe two demo apps that hide the transmission of private location data to the Internet by employing permission spreading via an implemented Service (see Figure 2). The app *Backdoor* requires the *access fine location* (GPS) permission, which could be justified by posing as GPS app that displays the current position. However, not detectable by the user the app also implements an Android Service that provides the GPS position to other apps. The second app – *TwitterApp* – has the *Internet* permission and could pose as a simple app for accessing Twitter, which again would not raise any suspicion during its installation. However, the user is not able to see that *TwitterApp* has malicious code that determines the current GPS position by calling the Service of app *Backdoor*. This GPS position is then posted to a Twitter account[9], which requires the *Internet* permission. Therefore, the app *TwitterApp* gains additional capabilities by calling a Service of another app and uses its own *Internet* permission to publish this information. Although the permission system is not directly circumvented (the app is still not able to get the GPS position without the Service of the second app), there are serious implications when analyzing this method in the context with malicious applications.

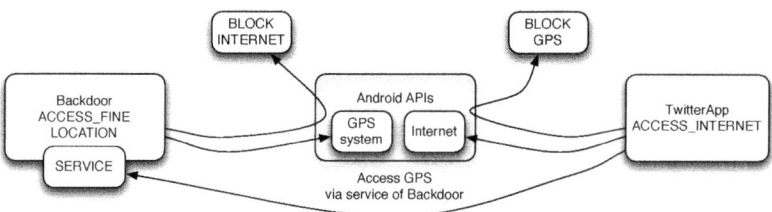

Fig. 2. The app *TwitterApp* uses the Service of *Backdoor* to determine the users's GPS position and submits it to Twitter via its own *Internet* permission

5.2 Implications for Malware

When taking a closer look at these two demo apps and the permission spreading method, we come to several conclusions:

Losing the Permission Context in the Android Market? The Android Market permission system is intended to support the user in her decision whether to trust an app before its installation.

---
[9] http://twitter.com/demolocator

Thereby, primarily the context in which multiple permissions are used and not only the permissions themselves, will alarm a user when inspecting a possible malicious application. However, exactly this context is lost in our presented attack, since permissions can be distributed over different apps and do not occur within the same context.

Trust in the Permission System? The Android permission system conveys a level of trust when an app is installed, since the system functionality can only be accessed when the appropriate permissions are available. However, when employing permission spreading, this trust leads to a wrong sense of security. In this case even an app without any permissions at all can gain additional functionality by calling functions within other apps that have these permissions.

Where Is the Malicious Code? Assuming, we have a malicious app that transmits private information information to an attacker without using permission spreading, then, firstly this app must declare all the required permissions (e.g., read contact data and Internet access) and secondly it must contain the malicious code that reads the private data and transmits it to the Internet. Such an app might raise suspicions due to the employed permissions and the lack of an adequate description or app use case that would necessitate these permissions. When analyzed thoroughly, the malicious activity could be detected by decompiling the application, capturing network traffic or employing other methods for detecting malware.

However, when employing permission spreading it is not necessary that the malicious code is contained within the app that has the permissions required for the malicious activity. For instance, app *Backdoor* only contains a Service that provides the GPS position, but not the malicious code that transmits this data to an attacker. Therefore, an arbitrary malware detection/analysis method would never detect the malicious activity when inspecting *Backdoor* only. In fact, *TwitterApp* carries the malicious code and uses *Backdoors's* permissions and Services to gain the information required for the malicious activity.

Furthermore, the app *TwitterApp* could come without any permission at all and just acquire the functionality through calling Services on multiple other apps (e.g., providing contact data, GPS position, Internet access).

Backdoors? Regardless of the malicious code's location and how the various available communication paths are employed, permission spreading still requires the installation of multiple apps by the smartphone users. At a first glance this might make the likelihood of a successful attack smaller. However, permission spreading could also be viewed as a classic backdoor that could either be injected or integrated on purpose into a popular application:

– Such a backdoor Service could be injected into existing source code that is not protected adequately. Since, the malicious code is not present but only the code required for providing certain information or functionalities, it might be difficult to detect such a backdoor – especially when communication side channels are employed (e.g. by writing data to a system log).

- The backdoor could be injected on purpose into a popular app by the company developing the app itself, by a developer that is involved in the development of an popular app or by a government[10].
- The backdoors could be integrated into multiple apps created by the same developer who then convinces the user to install more than one of the apps (e.g. by promoting add-ons, splitting functionality, additional levels for a game, by using a common API for multiple apps, by providing demo/full versions of applications, etc.).

6 Countermeasures

During the analysis of the permission spreading problem, we have also investigated countermeasures and implemented another demonstration app that focuses on the detection of a communication path between permission spreading apps. Concerning detection, we need to deal with the following question: *How can we detect malware that employs permission spreading?* The answer strongly depends on the employed communication method. We will give a short overview about the possible detection methods in the following sections.

6.1 Service Detection

Android Services are the simplest method for establishing a two-way communication path in Android. Services must be declared within the app manifest and get an identifier that is used when calling the Service. The detection therefore can be categorized into the detection of a service and the detection of a call to this service:

Detection of a Service

- **Android Market - by the User:** The Android market does not state which Services are provided within the application. Therefore, the user is not able to get information about the Services employed by an application.
- **Android Market - by Google:** Since the manifest of each app is readable, Google would be able to gain information about the employed Services within all market applications.
- **Android Smartphone - by the User:** The user is able to get information about running Services from the Android system. However, non-running Services are not displayed by the standard apps bundled with the Android system.
- **Android Smartphone - by an App (e.g., a virus scanner):** An arbitrary app without any permissions can query the Android `PackageManager` for installed applications. For each of these installed apps it is possible to list the declared Services. In addition the `ActivityManager` can be used to list the running Services.

---

[10] The possible attempt to catch Facebook account data in Tunisia is a good example for such an attack: `http://www.wired.com/threatlevel/2011/01/tunisia/`

Detection of Service Calls

Services are called by using Intents as parameters for the `startService()` or `bindService()` methods. The `bindService()` method enables the calling app to maintain a communication channel that is used for information exchange. The direct detection of such an Intent, the activation of a Service, or an established communication channel would require direct access to the Android system, which to our current knowledge is not possible without adding appropriate functions to the Android source code. At least for the `bindService()` method we have discovered a simple method, that allows us to determine when a Service is called and limit the possible apps that issued this call. We have also created a simple demo app (*ServiceBindDetection*) that can be installed as background Service and notifies the user whenever a Service is called by another application:

- The Android system Service allows an app to retrieve a list of all running Services (extracted via the `AcitivityManager`). In addition, for each Service the number of connected clients can be extracted. A client is connected when a *ServiceConnection* is maintained between a Service and the app that calls this Service.
- The detection app runs a loop in the background that in each iteration stores the running Services and their client count. Whenever the client count changes the user is notified within the Android notification bar.
- Whenever such a change occurs the detector could also get a list of currently running tasks (also via the `ActivityManager`) and thereby limit the possible callers[11]. By observing different calls to the same Service over time the possible perpetrators could be narrowed down.

We emphasize that this method does not work when `startService()` is used, since it does not maintain a communication channel and therefore does not list the calling app as connected client. Furthermore, we might miss certain calls when the duration of a `bindService()` `ServiceConnection` is smaller than the idle time of the Service checker loop.

6.2   Detection of Alternative Communication Paths

As described in Section 3, numerous ways for communication between applications exist. Acquiring information on interprocess communication other than via binding to Services turns out to be difficult. Information about Intents being sent would be valuable to detection of permission spreading. However, there does not seem to be a user-mode facility to obtain such information. Ongoing research will focus in this area.

Detection of communication via side channels like reading and writing to Content Providers seems to be even more difficult. Only in-depth analysis of the involved applications might yield satisfactory results.

---

[11] The most probable caller is the app that is currently running, however it cannot be ruled out that another app running in the background issues the call.

7 Conclusions and Outlook

The security of the Android permission system and the trust placed into the system is based on the assumptions that an app only has access to the functionality defined by the stated permissions and that all employed permission are displayed to the user within the same context (the app to be installed). As we show, these assumptions are not valid, since permissions can be spread over multiple apps that use arbitrary communication paths to gain functionality for which they do not have the appropriate permissions.

The intention of this work is to highlight the possible dangers and the wrong sense of security when trusting the permission system. Thereby, possible countermeasure range

- from making changes to the permission system including requiring permissions when using IPC between applications, or displaying communication interfaces prior to app installation,
- over implementing automatic detection systems within the Android Market, or performing an in-depth analysis of APK files,
- to shift the detection to the Android smartphone, by detecting communication events caused by permission spreading.

Addressing the detection on smartphones, we have presented a method to detect a covert communication channel involving Services. However, further investigations are necessary, since there is a large number of possible communication channels, ranging from documented IPC to not so obvious side channels.

References

1. Enck, W., Ongtang, M., McDaniel, P.: Understanding Android Security. IEEE Security and Privacy 7, 50–57 (2009)
2. Nauman, M., Khan, S., Zhang, X.: Apex: extending Android permission model and enforcement with user-defined runtime constraints. In: Proceedings of the 5th ACM Symposium on Information, Computer and Communications Security, pp. 328–332. ACM (2010)
3. Ongtang, M., McLaughlin, S., Enck, W., McDaniel, P.: Semantically Rich Application-Centric Security in Android. In: 2009 Annual Computer Security Applications Conference, pp. 340–349 (December 2009)
4. Shabtai, A., Fledel, Y., Kanonov, U., Elovici, Y., Dolev, S., Glezer, C.: Google Android: A Comprehensive Security Assessment. IEEE Security & Privacy Magazine 8(2), 35–44 (2010)
5. Toninelli, A., Montanari, R., Lassila, O., Khushraj, D.: What's on Users' Minds? Toward a Usable Smart Phone Security Model. IEEE Pervasive Computing 8(2), 32–39 (2009)
6. Vennon, T., Stroop, D.: Android Market: Threat Analysis of the Android Market (2010)

Private Pooling: A Privacy-Preserving Approach for Mobile Collaborative Sensing

Kevin Wiesner, Michael Dürr, and Markus Duchon

Mobile and Distributed Systems Group
Ludwig-Maximilian-Universität München
Oettingenstr. 67, 80538 Munich, Germany
firstname.lastname@ifi.lmu.de
http://www.mobile.ifi.uni-muenchen.de/

Abstract. Due to the emergence of embedded sensors in many mobile devices, mobile and people-centric sensing has become an interesting research field. A major aspect in this field is that quality and reliability of measurements highly depend on the device's position and sensing context. A sound level measurement, for instance, delivers highly differing values whether sensed from inside a pocket or while carried in a user's hand. Mobile collaborative sensing approaches try to overcome this problem by integrating several mobile devices as information sources in order to increase sensing accuracy. However, sharing data with other devices for collaborative sensing in return raises privacy concerns. By exchanging sensed values and context events, users might give away sensitive data, which should not be linkable to them. In this paper, we present a new mobile collaborative sensing protocol, *Private Pooling*, which protects the users' privacy by decoupling the data from its contributors in order to allow for anonymous aggregation of sensing information.

Keywords: Mobile Collaborative Sensing, Privacy, Ad hoc sharing.

1 Introduction

The recent development of mobile phones brought up various devices with embedded sensors. One popular example is the iPhone 4 that has a built-in gyrometer, accelerometer, proximity, and an ambient light sensor. In addition it has also integrated hardware for positioning and navigation (A-GPS, digital compass, Wi-Fi, Cellular) as well as for image and sound capturing (camera, microphone). The current Android reference [1] also supports various embedded sensors, for instance temperature and pressure sensors. It can be foreseen that upcoming mobile devices will come along with even more kinds of sensors. Possible future sensors might be environmental sensors (e.g. air pollution sensors), weather sensors (e.g. humidity sensors), as well as health sensors (e.g. heart rate sensors). Conceptual designs for smartphones with such a variety of sensing hardware already exist, for instance with the Nokia Eco Sensor Concept [17].

This development led to the research field of mobile or people-centric sensing networks (PCSN). In contrast to wireless sensor networks (WSN) where sensor

R. Prasad et al. (Eds.): MOBISEC 2011, LNICST 94, pp. 52–63, 2012.

are typically statically distributed, people-centric sensing (PCS) deals with mobile sensors embedded in user devices. In addition to mobility, we identified four main aspects in which PCSNs differ from WSNs:

1. Energy: In typical sensor networks, energy constraints are one of the main challenges [7]. However, as PCSN leverage user devices, energy is not as constrained as in typical WSNs. The life-time of static sensors is usually limited by the energy supply, whereas mobile phones can be recharged more easily. Thus, the challenge in PCSN is not the technical energy supply, but rather the willingness of users to share this limited resource. A possible solution to motivate users to share their resources is e.g. shown in [20].
2. Large-scale deployment: As more and more sensors will be built into regular mobile phones, there will be a huge amount of sensors that could be tapped for sensing applications. Large-scale sensor deployments are usually constrained by hardware costs, however, in PCSNs there are no such costs as users pay for it by buying mobile phones.
3. Sensor diversity: WSNs typically consist of a very homogeneous set of hardware, i.e., normally similar hardware is used throughout the network. In contrast, PCSNs consists of a very heterogeneous set of hardware as all different kinds of mobile devices could participate. This leads to differences in quality and accuracy of measurements that need to be coped with.
4. Privacy: As sensors are carried around by users, measurements reflect users' activities and locations. Thus, one main challenge of PCS is to conduct measurements and nevertheless preserve user privacy.

Another research field that builds upon embedded sensors are mobile context-aware systems, where built-in sensors of mobile devices are exploited to infer user context. For instance, many systems use a built-in accelerometer and gyrometer to determine the current user activity or setting [6,18]. This inferred information is then often used to adapt which and how information is presented, or to automate some processes with the goal to ease a user's life.

In either mentioned application field, quality and reliability of measurements highly depends on the device's sensing context, i.e., its position in relation to the information source [16]. That means, it is crucial where and how a mobile phone is carried by the user. The accelerometer reports different movement patterns when the phone is carried around in the pants pocket, compared to measurements conducted while carried in the handbag. A sound level measurement, for instance, delivers highly differing values if sensed with a mobile phone inside a bag compared to results where the device was carried in the hand. A context-aware application could for instance reason that a person is in a quiet room, while actually being in a crowded public space, just because the phone's microphone is in a pocket and sound patterns hence differ significantly.

Mobile collaborative sensing approaches try to overcome this problem. Instead of relying on one single value that might deliver distorted measurements, those concepts integrate several mobile sensing devices as information sources. The initiator of the request receives the collected information, and can then reach a decision taking the different sensed valued into account. If several phones detect the same event, the probability that the event happened increases. If enough measurements are available, it is possible to statistically eliminate some of the measurements. Thereby, more robust and reliable results can be achieved.

A very important aspect of mobile sensing is the protection of privacy, as participating users might reveal sensitive information by providing their phone-based measurements. Collaborative sensing approaches even intensify these privacy concerns, as information is shared with anybody that requests it and thus provided data could be misused. To prevent this, shared information about sensed values or detected context events should be exchanged in an anonymized way. For ad hoc sharing of mobile sensors, user intervention is not reasonable, thus user privacy needs to be automatically adhered.

We present a new mobile collaborative sensing method, *Private Pooling*, which describes an approach to exchange sensed information anonymously by incorporating concepts from multi-party computation protocols. Private Pooling allows for aggregated sensing and collaborative context information exchange among co-located phones, without revealing the link between information and its contributor.

The remainder of this paper is structured as follows. In Section 2, a motivating example is given, and Section 3 outlines some fundamentals. Section 4 then describes our concept and proposed solution, followed by a discussion in Section 5. Section 6 describes the current prototype implementation, and Section 7 gives an overview of related work and projects. Finally, in Section 8 we summarize our results and conclude with an outlook on future work.

2   Motivation

In this section, we want to emphasize the need for a privacy-preserving protocol in the field of mobile collaborative sensing. A first question could be, why we aim for a protocol instead of letting users decide which information they want to share. There exist two reasons why we favor automated sharing: (1) Mobile collaborative sensing is supposed to be an ubiquitous technology, that should merge in the background. If people are required to actively accept sharing their sensor data, users could get annoyed by responding to these requests. The benefit of mobile collaborative sensing increases with the number of participants. It should be as easy as possible to participate in order to foster a large-scale deployment (cf. [12]). (2) The response time of participating phones should be very short, so that a requesting user can use the gathered information with only a minimum delay. Waiting for users to respond to requests whether they would like to share their data or not could protract the whole process. However, unobtrusive sharing of information also leads to a loss of control which information is shared with whom and when. Attackers could exploit unintentionally shared data. There are several use cases, where this loss of control could be problematic. In the following, we focus on one short illustrative example:

Alice, Bob, Carol, and Mallory are all at the same train station, and reside within radio range of Mallory, so that they can communicate directly with him. An application on Mallory's phone now wants to use collaborative sensing to infer the current context. Even though the term "context" typically refers to a description of "a situation and the environment a device or user is in" [19], in this example for the sake of simplicity we assume that context is a textual description of the environment the user is in. Further, we assume that the context recognition is part of the application, and that the application provides some context model, as our approach focuses on information exchange rather than on information reasoning. A possible model delivered could consist of a determined context in combination with a certainty that this context is correct ({context: certainty}). The application sends out a context aggregation requests, and gathers responses of the others. Bob and Carol might both return a simple model containing {train station : 100%}. Since Alice likes jewelry, she often visits a near-by jewelry store, which is well-known for its first-class gem. Her phone already learned that being in that area at that time often means that she is in the jewelry store, and thus her context response could look like the following: {train station: 82%, jewelry: 18%}. Alice's response contains highly sensitive information, which she does not want Mallory to know about. The application's privacy problem could thus lead to a real-world security risk for Alice, if Mallory knew that this information was contributed by her. Mallory could for instance reason that Alice is wealthy and use this information with criminal intent.

It becomes apparent that there is a strong need for privacy when collaborative sensing is applied. One way to solve this would be to blur or reduce the data that is shared. Alice could only send {train station: 82%} so that no sensitive data would leak. This raises two problems: First, this approach reduces the quality of data and therefore the usefulness might decrease as well. Second, it is impossible to automatically infer which parts are sensitive and which are not. Thus, our approach focuses on separating the data from the participants, i.e., shared data must not be linkable to the user who shared it. In our example, Mallory would receive a result {train station: 94%, jewelry: 6%}, but could not determine which user contributed which part.

3   Preliminaries

As pointed out in the previous section, the goal of our approach is to allow for sensor data exchange over insecure wireless connections without revealing individually contributed information. Users should be able to contribute their information without any privacy concerns, as nobody should be able to discover which part of the collected data was contributed by whom.

In the following, we first identify the requirements that a solution should fulfill. Subsequently, we shortly explain which concepts our approach is derived from.

3.1   Requirements

In this section we outline which requirements an optimal solution should fulfill. Mobile collaborative sensing should be employed for more accurate and robust sensing. Thus it should leverage the exchange of sensor data without harming users by revealing sensitive information. In this setting, we identified three main requirements:

Privacy-Preserving: The protection of the participants' privacy is the main requisite, as already motivated before.

Decentralization: If users are in a setting with no or only limited connectivity, e.g. in a building, it should still be possible to receive the data of surrounding mobile phones, without depending on a central backend server.

Lossless Information Exchange: An optimal solution should be lossless, that means available data should be completely shared in best quality.

3.2   Background

Our concept was designed with the previous requirements in mind. One area where the mentioned aspects are already considered for data exchange is the field of secure multi-party computation (MPC). The aim of MPC is to provide secure solutions for joint computation with private inputs from multiple parties, typically by leveraging public-key cryptography. There exist several MPC approaches including approaches for secure joint computation, joint signatures, elections over the Internet, and private database access [10,9]. Further approaches are outlined in Section 7. Our concept of *Private Pooling* is based on a solution for the so-called "millionaires' problem". It originally refers to a situation where two millionaires want to know who of them is richer without revealing their wealth to their counterpart. Yao [21] proposed a solution for this problem. For our concept an adapted version of this problem is used as a basis.

Grosskreutz, Lemmen, and Rüping [11] describe a simple solution to allow multiple millionaires to find out how much money they own all together without revealing the individual's wealth. In Figure 1, the concept of this approach is illustrated. The first participant generates a random number (rnd) between 0 and M, where M is the upper bound of the outcome. Then he adds the value of his assets ($v1$), and sends this combined value modulo M to participant 2. As the random number is unknown to participant 2, he cannot determine the actual wealth of participant 1. From now on, the participants simply add their wealth to the aggregated sum (modulo M). The last participant returns the value, consisting of the random number and all participants' values, to participant 1, who can then subtract the initially added random value (rnd) and finally receives the value of the total wealth of all millionaires. This approach obviously only works in the semi-honest model, that means, participants try to find out as much as possible but all stick to the protocol. It does not work in case of a shared medium, as every communication can be overheard by all users.

In the following section, we outline our adaptation of this protocol to the mobile and collaborative sensing scenario.

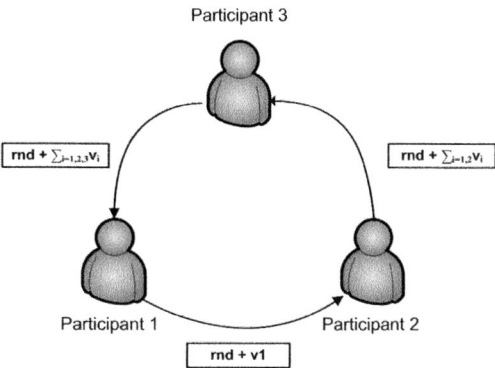

Fig. 1. Secure multi-party sum computation

4   Concept

Our proposed concept, *Private Pooling*, is a protocol that enables privacy-preserving information exchange over wireless connections. This is reached by establishing secured circular communication and decoupling shared information from its contributor.

Our main contribution is the concept of secured circular communication, which refers to a circular communication order of participating users in such a way that communication between two users cannot be eavesdropped by other users in the vicinity. This adaptation of the aforementioned millionaires example allows us to enforce a certain, specified communication order, even in the mobile and wireless scenario. In combination with additional randomized data, it can be ensured that shared information cannot be backtracked to the users that contributed it, as in the aforementioned millionaires example.

Private Pooling consists of three phases: (1) Sharing Request Phase, (2) Sharing Response Phase, and (3) Information Aggregation Phase. In the following we explain all phases in detail:

(1) *Sharing Request Phase*: A collaborative sensing activity starts with the Sharing Request Phase (see Figure 2a). A user that wants to gather information of surrounding users (in the following called requester), broadcasts an Information Sharing Request (*ISReq*). The *ISReq* contains the type of information that the requester wants to collect, and thus notifies all users in his vicinity about his interest. Further, the *ISReq* specifies the minimum number of participants (*Min-Collaborators*). The reason for this is to notify users about the privacy level of this sharing request (*ISReq*). The more users participate, the more difficult it is to link some data to a specific user. The usage of *MinCollaborators* will be explained later on. Collected information by only two other users can be more easily backtracked to the contributing users than if it is an information pool gathered by 20 users. Further, the *ISReq* contains the requester's public key to enable secure communication at a later phase.

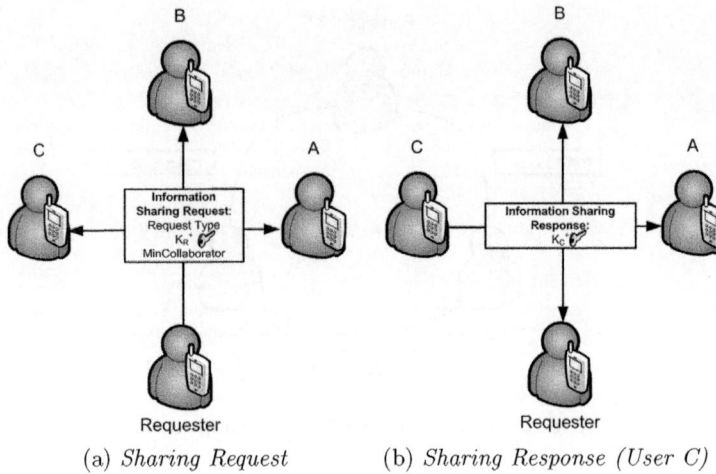

(a) *Sharing Request*   (b) *Sharing Response (User C)*

Fig. 2. Communication flow for initial phases

(2) *Sharing Response Phase*: In the second step, all users willing to participate, respond to the *ISReq* with an Information Sharing Response (*ISRes*). By sending an *ISRes* participants indicate that they are willing to contribute their sensed data to the data collection initiated by the requester (Figure 2b shows this process for user C only, but A and B need to respond in a similar way). The public key of the user is included in this *ISRes*, so that possible future messages addressed to him can be encrypted. As the *ISReq*, the *ISRes* is broadcasted too. Thereby, other participating users learn which contributors are within their reach, and can store their public keys so that they can send encrypted messages in the future. Users that are not in reach, are not able to exchange keys, and consequently will not be able to communicate directly, even if they happen to be within radio range later on.

(3) *Information Aggregation Phase*: This last step is the actual collaborative part where sensed data is collected. If a requester receives enough *ISRes*, i.e. if $\sum$ ISRes >= MinCollaborators, the requester can start the Information Aggregation Phase by sending out an Information Aggregation Request (*IAReq*). An *IAReq* consists of two main parts: the sensed data and aggregator information. The latter contains a list of users that are willing to participate (*scheduled participants*) signed with the private key (K_R) of the requester and an unsigned list of users that already contributed their data to this collection (*completed participants*). By signing the list of scheduled participants, each user can verify that this list was not modified by any other user than the requester. However, the second list has to be updated by each user, and could be also subject to forgeries. But in contrast to the former, the completed participant list cannot be modified in such a way that the modifying user would benefit from his changes himself. For the sake of simplicity, we will use the term *aggregator list* in the following for the remaining users that still need to participate ($P_{\text{scheduled}} \setminus P_{\text{completed}}$).

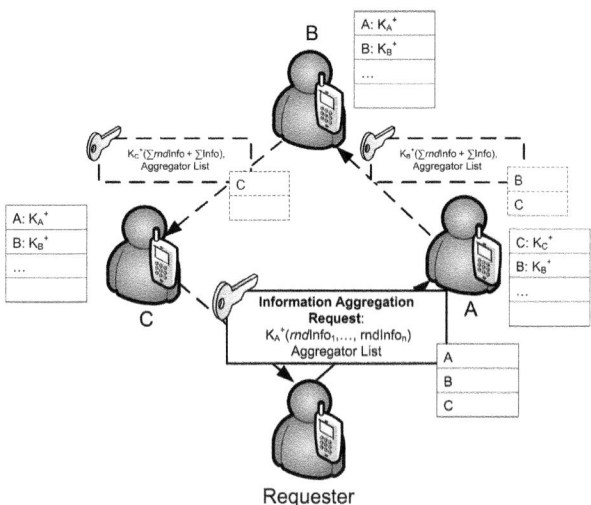

Fig. 3. Communication flow in the *Information Aggregation Phase*

Since the requester's direct successor knows that he is the first participant (because he receives the *IAReq* from the same user that broadcasted the *ISReq*), the requester does not contribute any real data in order to protect his privacy. Instead he generates n random data sets. Random data sets are generated from previously collected data sets of other users mixed with own former data, in order to provide highly realistic but uncritical data. If the requester, for instance, already participated in k previous sharing processes, he received d_k data sets each time (depending on his position in this process). Combined with p own data sets, he posses $p + \sum_{i=1}^{k} d_i$ data sets, from which he randomly chooses n to forward to the first participant. In case of a collection of temperature data, this could look like the following: $\{14\,^\circ\text{C}, 15\,^\circ\text{C}, 23\,^\circ\text{C}\}$.

Subsequently, one user in the aggregator list is chosen as the first recipient, i.e. in our example in Figure 3 user A. The data is then encrypted with A's public key, and finally the *IAReq* is sent. As user A is the first participating user, A first checks whether the previously announced number of *MinCollaborators* is less or equal to the length of the received aggregator list. *MinCollaborators* specifies the minimum number of participants, and thus is a privacy indicator. If the length of the aggregator list does not comply with this number, the users privacy could be comprised, and therefore the *IAReq* is rejected. If the *IAReq* complies with the previously broadcasted *ISReq*, A selects one of the users (from which he received an *ISRes*) from the aggregator list as successor. The selection of the user is random, that means, there is no predetermined order in which the data is collected.

User A adds his sensed data, e.g. a temperature of $18\,^\circ\text{C}$, to the received data from the requesters, and forwards the *IAReq* to the next participant, in our case user B. B would then remove his predecessor's (in this case A's) entry from the aggregator list (that means, he adds A to the list of completed participants), and would receive the following data: $\{14\,°C, 15\,°C, 18\,°C, 23\,°C\}$. The following users add their sensor data as well, and forward the data collection to a remaining participant as long as the aggregator list consists of more than only their own entry. If no more other entries exist in the aggregator list, the last user (here: user C) adds his data to the collected data and forwards it back to the requester using the requester's public key. The last step of the protocol is the requester's removal of the initially added random data sets and the contribution of his own data, which leads to the real data collection. Assuming that the requester receives $\{14\,°C, 15\,°C, 18\,°C, 20\,°C, 20\,°C, 23\,°C\}$ from user C and measured a temperature of $17\,°C$ himself, this would lead to the following final data set: $\{17\,°C, 18\,°C, 20\,°C, 20\,°C\}$.

In a further optional step, the requester could broadcast the collected data to all participants, as only the initiator knows the result. However, this might not be necessary in many cases, as the participating users might not be interested in the data at that moment. For that reason, the default is that collected information is not broadcasted. If participants are interested in receiving the aggregated information, they have to indicate it by setting a *BroadcastFlag*. In that way, the initiator knows whether a broadcast is necessary or not.

During this phase, it might happen that a user has not received any public key of the remaining users on the aggregator list during the *Sharing Response Phase*. Or the remaining participants are not in transmission range in the *Information Aggregation Phase* and cannot be reached anymore. In either case, this participant is not able to forward the collected data and sends an negative acknowledgement (*NACK*) to his predecessor. The predecessor then tries to forward the *IAReq* to another user on the aggregator list. If no other user is available, the *NACK* is further forwarded to preceding users until a successful communication order can be found or the requester receives the *NACK*. In the latter case, the requester simply triggers a new *ISReq* to overcome this problem.

5   Analysis and Discussion

First, we review the communication complexity of our approach: A naive approach where participants directly reply to the initiator's request requires n messages to gather measurements of n participants (*ISReq+(n-1)ISRes*). Our approach in contrast requires $2n$ messages, as it first establishes a secured circular communication before actual measurements are transmitted. Even though it is a considerable increase, it remains scalable due to its linear complexity ($O(n)$). The communication overhead arises mainly from the circular data collection. Instead of sending each measurement directly to the requester, measurement values are collected in a specified order, which leads to an increase of utilized data bits. For n participants and m random values, this leads to $nm + \frac{n(n-1)}{2}$ values (the first message contains m values, the following $m+1$ values, etc.) that are sent around, compared to $n-1$ values in the naive approach.

Second, we outline some shortcomings and limitations of our approach: Our protocol is designed for the semi-honest model, as the approach outlined in

Section 3.2. If two malicious participants collaborate and exchange information they received from others, it could be possible to determine the input of individuals. In our example (Figure 3), participant A and C could exchange and compare the received information and thus extract B's input. Another problem in this context could arise, if a requester simulates n users and responds to his own request with *(n-1) ISRes* messages. Other users would then believe that there are *(n-1)* users participating, even though they might be the only real contributor. This would allow the requester to simply extract the contributed data. In the current version, we do not address those problems, as we focused on providing anonymity for the use case of participants sticking to the protocol. Our concept tries to avoid that sensitive information could be leaked even without active involvement of malicious users. However, in our future work we intend to extend this scenario for the malicious model as well.

6   Implementation

We built a first prototype as a proof of concept. Therefore, we developed an Android application that provides some sensor data (i.e. temperature readings) and collects data from co-located phones. For discovering nearby phones and the communication between those, the *haggle* API [2] was utilized. Haggle provides a network architecture for opportunistic communication and enables content exchange based on interests with a publish/subscribe messaging system. Thus, *ISReq* and *ISRes* can be broadcasted, as all phones register to follow the *ISChannel*. Messages in the information aggregation phase are sent using individual channels for each user. To secure communication, we used public key encryption provided through the standard android reference. In our future work we plan on using this prototype to evaluate our proposed protocol in empirical studies.

7   Related Work

As this paper combines aspects from mobile collaborative sensing as well as secure multi-party computation, relevant related work can also be found in these two research areas.

Mobile collaborative sensing is an approach which arose from mobile sensing research, often named people-centric sensing [5]. There has been a lot of work done on applications or systems that leverage embedded sensors of mobile phones [15,8,13,3], however most of the work focuses on individually sensed measurements. The concept of combining multiple information sources and co-operating sensors originates from static sensor networks, where approaches such as sensor fusion and aggregation have already been intensively studied. In the field of mobile phone-based sensing, only few works have examined collaborative sensor usage. One approach was proposed by Miluzzo et al. [14]: Their collaborative reasoning framework "Darwin Phones" allows for pooling of evolved models with neighboring phones, as well as collaborative inference for jointly observed events. Thereby, the authors want to improve the model development and aim on making the inference more robust. Another example is presented by Bao and Choudhury [4]: They designed a phone application that collaboratively recognizes socially interesting events and automatically clusters video recordings around these detected events.

However, the mentioned mobile collaborative sensing approaches have not really addressed privacy and security issues as highlighted in this paper. Therefore, we proposed a protocol that is based on secure multi-party computation (MPC) concepts. MPC is a research field of cryptography, with the aim to enable secure execution of distributed computing tasks, without the need for a trusted third party. Therefore it usually leverages public-key cryptography. Yao [21] proposed a solution for the so-called "millionaires' problem", where two millionaires want to know who of them is richer without revealing their wealth to their counterpart. Other MPC concepts address for instance the issue of sharing a secret among multiple parties, computing random bit choices by multiple parties, or oblivious transfers. An overview of MPC problems and applications can be found in [9].

To the best of our knowledge, the utilization of multi-party approaches for wireless ad-hoc communication in the field of mobile collaborative sensing has not been studied yet, and thus makes a unique contribution to this field.

8   Conclusion and Future Work

This paper explores a new approach for mobile collaborative sensing. The proposed "Private Pooling" protocol leverages anonymous ad-hoc sensing data collection of co-located mobile phone users. In order to achieve a secure and privacy-preserving information exchange, our concept is based on a multi-party computation approach and prevents others from linking shared data to the person that shared this data. Thereby, we allow for a privacy-preserving information exchange without reducing quality of contributed data. We also discussed shortcomings of the current protocol and outlined our current implementation.

The next steps will be to further improve our protocol to make it more robust (e.g. also in case of the malicious model) and to conduct empirical studies using our prototype.

References

1. Android reference (December 2010), http://developer.android.com/reference/
2. haggle - A content-centric network architecture for opportunistic communication (2010), http://code.google.com/p/haggle/
3. Abdelzaher, T., Anokwa, Y., Boda, P., Burke, J., Estrin, D., Guibas, L., Kansal, A., Madden, S., Reich, J.: Mobiscopes for human spaces. IEEE Pervasive Computing 6, 20–29 (2007)
4. Bao, X., Choudhury, R.R.: Movi: mobile phone based video highlights via collaborative sensing, pp. 357–370 (2010)
5. Campbell, A.T., Eisenman, S.B., Lane, N.D., Miluzzo, E., Peterson, R.A.: People-centric urban sensing. In: Proceedings of WICON 2006, p. 18. ACM, New York (2006)

6. Choudhury, T., Borriello, G., Consolvo, S., Haehnel, D., Harrison, B., Hemingway, B., Hightower, J., Klasnja, P.P., Koscher, K., LaMarca, A., Landay, J.A., LeGrand, L., Lester, J., Rahimi, A., Rea, A., Wyatt, D.: The mobile sensing platform: An embedded activity recognition system. IEEE Pervasive Computing 7, 32–41 (2008)
7. Dargie, W., Poellabauer, C.: Fundamentals of Wireless Sensor Networks: Theory and Practice. Wiley (2010)
8. Das, T., Mohan, P., Padmanabhan, V.N., Ramjee, R., Sharma, A.: Prism: platform for remote sensing using smartphones. In: Proceedings of MobiSys 2010, pp. 63–76. ACM, New York (2010)
9. Du, W., Atallah, M.J.: Secure multi-party computation problems and their applications: a review and open problems, pp. 13–22 (2001)
10. Goldwasser, S.: Multi party computations: past and present. In: Proceedings of PODC 1997, pp. 1–6. ACM, New York (1997)
11. Grosskreutz, H., Lemmen, B., Rüping, S.: Privacy-preserving data-mining. Informatik-Spektrum 33, 380–383 (2010)
12. Lane, N.D., Eisenman, S.B., Musolesi, M., Miluzzo, E., Campbell, A.T.: Urban sensing systems: opportunistic or participatory? In: Proceedings of HotMobile 2008, pp. 11–16. ACM, New York (2008)
13. Lu, H., Pan, W., Lane, N.D., Choudhury, T., Campbell, A.T.: Soundsense: scalable sound sensing for people-centric applications on mobile phones. In: Proceedings of MobiSys 2009, pp. 165–178. ACM, New York (2009)
14. Miluzzo, E., Cornelius, C.T., Ramaswamy, A., Choudhury, T., Liu, Z., Campbell, A.T.: Darwin phones: the evolution of sensing and inference on mobile phones. In: Proceedings of MobiSys 2010, pp. 5–20. ACM, New York (2010)
15. Miluzzo, E., Lane, N., Eisenman, S., Campbell, A.: CenceMe: Injecting Sensing Presence into Social Networking Applications, pp. 1–28 (2007)
16. Miluzzo, E., Papandrea, M., Lane, N.D., Lu, H., Campbell, A.T.: Pocket, Bag, Hand, etc.-Automatically Detecting Phone Context through Discovery, http://www.cs.dartmouth.edu/miluzzo/papers/miluzzo-phonesense10.pdf
17. Nokia. Nokia eco sensor concept, http://www.nokia.com/environment/devices-and-services/devices-and-accessories/future-concepts/eco-sensor-concept
18. Raento, M., Oulasvirta, A., Petit, R., Toivonen, H.: Contextphone: A prototyping platform for context-aware mobile applications. IEEE Pervasive Computing 4, 51–59 (2005)
19. Schmidt, A., Beigl, M., Gellersen, H.-W.: There is more to context than location. Computers & Graphics 23(6), 893–901 (1999)
20. Teske, H., Furthmüller, J., Waldhorst, O.P.: A Resilient and Energy-saving Incentive System for Resource Sharing in MANETs. In: Proceedings of KiVS 2011, Dagstuhl, Germany. OASIcs, vol. 17, pp. 109–120 (2011)
21. Yao, A.C.: Protocols for secure computations. In: Annual IEEE Symposium on Foundations of Computer Science, pp. 160–164 (1982)

Agent Based Middleware for Maintaining User Privacy in IPTV Recommender Services

Ahmed M. Elmisery and Dmitri Botvich

Telecommunications Software & Systems Group
Waterford Institute of Technology, Waterford, Ireland

Abstract. Recommender services that are currently used by IPTV providers help customers to find suitable content according to their preferences and increase overall content sales. Such systems provide competitive advantage over other IPTV providers and improve the overall performance of the current systems by building up an overlay that increases content availability, prioritization and distribution that is based on users' interests. Current implementations are mostly centralized recommender service (CRS) where the information about the users' profiles is stored in a single server. This type of design poses a severe privacy hazard, since the users' profiles are fully under the control of the CRS and the users have to fully trust the CRS to keep their profiles private. In this paper, we present our approach to build a private centralized recommender service (PCRS) using collaborative filtering techniques and an agent based middleware for private recommendations (AMPR). The AMPR ensures user profile privacy in the recommendation process. We introduce two obfuscation algorithms embedded in the AMPR that protect users' profile privacy as well as preserve the aggregates in the dataset in order to maximize the usability of information for accurate recommendations. Using these algorithms provides the user complete control on the privacy of his personal profile. We also provide an IPTV network scenario that uses AMPR and its evaluations.

Keywords: Privacy, Clustering, IPTV Networks, Recommender System, Multi-Agent Systems.

1   Introduction

Internet protocol television (IPTV) is one of the most fast growing services in ICT; it broadcasts multimedia content in digital format via broadband internet networks using IP packet switched network infrastructure. Differently from conventional television, IPTV allows an interactive navigation of the available items [1]. IPTV providers employ automated recommender services by collecting information about user preferences for different items to create a user profile. The preferences of a user in the past can help the recommender service to predict other items that might be interested for him in the future.

Collaborative filtering (CF) technique is utilized for recommendation purposes as one of the main tools for recommender systems. CF is based on the assumption that

R. Prasad et al. (Eds.): MOBISEC 2011, LNICST 94, pp. 64–75, 2012.

people with similar tastes prefer the same items. In order to generate recommendations, CF cluster users with the highest similarity in their interests, then dynamic recommendations are then served to them as a function of aggregate cluster interests. Thus, the more the users reveal information about their preferences, the more accurate recommendations provided to them. However at the same time the more information is revealed to the recommender service about the user profile, the lower user privacy levels can be guaranteed. This trade-off acts as a requirement when designing a recommender service using CF technique. Privacy aware users refrain from providing accurate information because of their fears of personal safety and the lack of laws that govern the use and distribution of these data. Most service providers would try their best to keep the privacy of their users. But occasionally, when they are facing bankruptcy, they might sell it to third parties in exchange of financial benefits. In the other side, many service providers might violate users' privacy for their own commercial benefits. Based on a survey results in [2, 3] the users might leave a service provider because of privacy concerns. The information collected by recommender service breaches the privacy of the users in two levels.

1. The real identity of the user is available to a central server. That server can associate the user profile which contains his private information to his real identity. This is an obvious privacy breach, considering that a user does not want to reveal the link between his real identity and his profile, yet he wants to use the service in that server.
2. If the user is not known to the server, the server can try to de-anonymize the user identity by correlating the information contained in the user profile and some information obtained from other databases [4].

In this paper we proposed an agent based middleware for private recommendation (AMPR) that bear in mind privacy issues related to the utilization of collaborative filtering technique in recommender service and allow sharing data among different users in the network. We also present two obfuscation algorithms that protect the user privacy and preserve the aggregates in the dataset to maximize the usability of information in order to get accurate recommendations. Using these algorithms, gives the user a complete control on his personal profile, so he can make sure that the data does not leaves his side until it is properly desensitized. In the rest of this paper we will generically refer to news programs, movies and video on demand contents as Items. Section 2 describes some related work. In Section 3 we introduce our private centralized recommender service scenario in IPTV network. In Section 4 we introduce the proposed obfuscation algorithms used in our framework. Section 5 describes some experiments and results based on obfuscation algorithms for IPTV network. Section 6 includes the conclusion and future work.

2   Related Work

The majority of the literature addresses the problem of privacy for recommender services based on collaborative filtering technique, Due to it is a potential source of leakage of private information shared by the users as shown in [5]. In [6] it is proposed a theoretical framework to preserve the privacy of customers and the commercial interests of merchants. Their system is a hybrid recommender that uses secure two party protocols and public key infrastructure to achieve the desired goals. In [7, 8] it is proposed a privacy preserving approach based on peer to peer techniques using users' communities, where the community will have a aggregate user profile representing the group as whole and not individual users. Personal information will be encrypted and the communication will be between individual users and not servers. Thus, the recommendations will be generated at client side. In [9, 10] it is suggest another method for privacy preserving on centralized recommender systems by adding uncertainty to the data by using a randomized perturbation technique while attempting to make sure that necessary statistical aggregates such as mean don't get disturbed much. Hence, the server has no knowledge about true values of individual rating profiles for each user. They demonstrate that this method does not decrease essentially the obtained accuracy of the results. Recent research work [11, 12] pointed out that these techniques don't provide levels of privacy as it was previously thought. In [12] it is pointed out that arbitrary randomization is not safe because it is easy to breach the privacy protection it offers. They proposed a random matrix based spectral filtering techniques to recover the original data from perturbed data. Their experiments revealed that in many cases random perturbation techniques preserve very little privacy. Similar limitations were detailed in [11].

3   Problem Formulation

3.1   System Model

We consider a system where PCRS is implemented as a third-party service that makes recommendations by consolidating the profiles received from multiple users. Each user has a set top box (STB) that stores his profile and host AMPR at his side. As shown in fig 1, the parties involved are the users, and the PCRS. We assume that PCRS follow the semi-honest adversary model, which is realistic assumption because the PCRS provider needs to accomplish some business goals and increase his revenues. Moreover, we assume the communication links between parties are secured by existing techniques. An IPTV provider uses this business model to reduce the required computational power, expenses or expertise to maintain an internal recommender service.

3.2   Design Goals

There are two requirements should be satisfied in the previous system model:

— IPTV providers care about the privacy of their catalogue which is considered an asset for their business. In the meantime they are willing to offer real users' ratings for different masked items to offer better recommendations for their users and increase their revenues.
— In the other side, privacy aware users worry about the privacy of their profiles, as sending their real ratings harm their privacy.

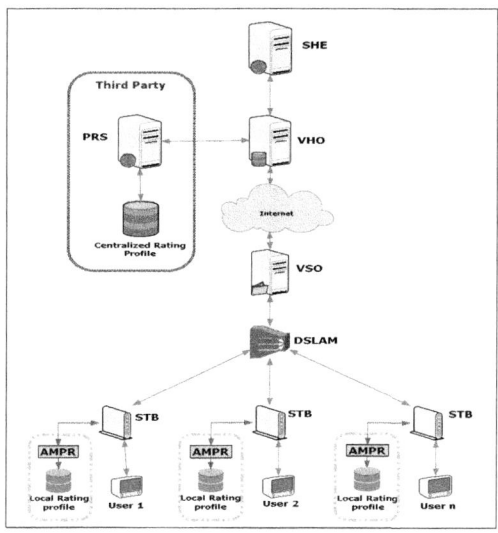

Fig. 1. Illustration of proposed combined IPTV Network

The AMPR employs two obfuscation algorithms that provide the users the required privacy level before submitting the profiles to the PCRS. Note that, we alleviate the user identity problems by using anonymous pseudonyms identities for users.

3.3   Threat Model

In this paper, AMPR provides a defence mechanism against the threat model proposed in [13] where the attacker colludes with some users inside the network to obtain some partial information about the process used to obfuscate the data and/or some of the original data items themselves. The attacker can then use this partial information for the reverse engineering of the entire data set.

4   Solution

In the next sections, we will present our proposed framework for preseving the privacy of customers' profiles show in fig 2.

4.1   PCRS Components

As show in fig 2, PCRS maintains a set data stores. The first data store is the masked catalogue of items that have been hashed using IPTV provider key or a group key. The second data store is the obfuscated users' profiles which contain users' pseudonyms and their obfuscated ratings and finally a peer cache which is an updated database about peers participated in previous recommendations formulation. The peer cache is updated from peer list database at client side. The PCRS communicates with the user through a manager unit. Finally, the clustering manager is the entity responsible for building recommendations model based upon the obfuscated ratings database.

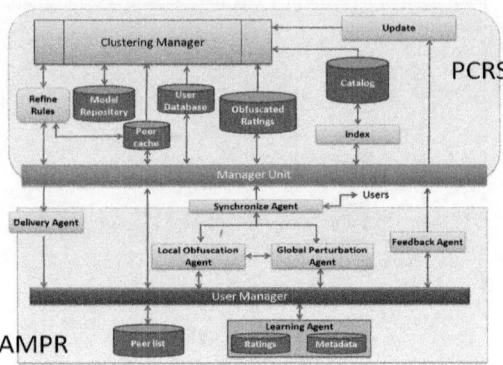

Fig. 2. PCRS framework

4.2   AMPR Components

The AMPR in the user side consists of different co-operative agents. Learning agent captures user preferences about items explicitly or implicitly to build a rating table and meta-data table. The local obfuscation agent implements CBT obfuscation algorithm to achieve user privacy while sharing the data with other users or the system. The global perturbation agent executes G-algorithm on the locally obfuscated collected profiles. These algorithms act as wrappers that obfuscate items' ratings before they are fed into the PCRS. Since the database is dynamic in nature, the local obfuscation agent desensitizes the updated data periodically, then synchronize agent send it to other users and PCRS. So the recommendations are made on the most recent ratings. More details about the recommendation process described in the next sub-section.

4.3   The Recommendation Process

The recommendation process based on the two stage obfuscation algorithms can be summarized as following more details can be found in [14]. The target user broadcasts message to other users in the IPTV network to request starting the recommendations process or update their centralized rating profiles stored at PCRS. The individual users who decided to participate in that process use the local obfuscation agent to perform *CBT* algorithm of their local rating profiles. They agree on same parameters, and then they submit their locally obfuscated profiles to the requester. The target user instructs his obfuscation agent to start *G* algorithm on the collected locally obfuscated profiles. After finishing the previous step, the target user submits all profiles to PCRS in order to receive recommendations.

5   Proposed Algorithms

In the next sub-sections, we provide two different algorithms that used by our agents to obfuscate the user profile in a way that secure his ratings in the un-trusted PCRS with minimum loss of accuracy. In our framework, each user has two datasets representing his/her profile. First one is the local rating profile which is perturbed before merging it with similar users' profiles that rare willing to collaborate with him as part of the recommendation process. The second one is the centralized rating profile which is the output of the two obfuscation algorithms where the user can get recommendation directly from the PCRS based on it. We perform experiments on real datasets to illustrate the applicability of our algorithms and the privacy and accuracy levels achieved using them.

5.1   Local Obfuscation Using CBT Algorithm

We propose a new obfuscation algorithm called clustering based transformation (CBT) that have been designed especially for the sparse data problem in user profile. It is inspired from the block cipher idea in [15]. We present a new technique to build a transformation lookup table (TLUT) using clustering technique then approximate each point in the data set to the nearest representative value in the TLUT (the core-point for the cluster it belong to) with the help of similarity measures. The output of our obfuscation algorithm should satisfy two requirements:

1. Reconstructing the original data from the obfuscated data should be difficult, in order to preserve privacy.
2. Preserve the similarity between data to achieve accurate results.

We use local learning analysis (*LLA*) clustering method proposed in [16] to create the *TLUT*. It is important to attain an optimized *TLUT* because the quality of the *TLUT* obviously affects the performance of the transformation. *LLA* builds an initial *TLUT* and repeats the iteration till two conditions satisfied:

1. The distance function $d(x, c_i)$ between a point x and its corresponding value (core-point) c_i is minimized.
2. The distortion function between each dataset and its nearest value (core-point) becomes smaller than a given threshold.

CBT algorithm consists of following steps:

1. The user ratings stored as dataset D of c rows, where each row is sequence of fields $X = x_1 \, x_2 \, x_3 \ldots\ldots x_m$.
2. User ratings dataset D is portioned into $D_1 \, D_2 \, D_3 \ldots\ldots D_n$ datasets of length L , if total number of attributes in original is not perfectly divisible by L then extra attributes is added with zero value which does not affect the result and later it is removed at step 5.
3. Generate *TLUT* using *LLA* algorithm, *LLA* takes Gaussian Influence function as the similarity measure. Influence function between two data points x_i and x_j is given as

$$f_{Gauss}^{x_i}(x_j) = e^{-\frac{d(x_i, x_j)^2}{2\sigma^2}} \tag{1}$$

While the field function for a candidate core-point given by:

$$f^{D}_{Gauss}(x_j) = \sum_{s=1}^{k} e^{-\frac{d(x_j,x_{is})^2}{2\sigma^2}} \qquad (2)$$

Clustering is performed on each dataset D_i, resulting to k clusters $C_{i1}, C_{i2}, C_{i3}, ..., C_{ik}$ and each cluster is represented by its core-points, i.e. core-point of jth cluster of ith dataset is $(C_{ij}) = \{c_1, c_2, c_3, ... c_L\}$. Every single row portion falls in exactly one cluster. And The TLUT = (core-point (C_{i1}), core-point(C_{i2}), core-point(C_{i3}).., core-point (C_{ik}))

4. Each dataset D_i is transformed into new dataset $D_i{'}$ using generated TLUT, each portion $Y_i = x_{(i-1)L+1} x_{(i-1)L+2} x_{(i-1)L+3} .... x_{iL}$ replaced by the nearest cluster core-point Z_i = core-point (C_{ij}) in which it falls.

$$Y_i \xrightarrow{transofmred} Z_i$$

5. The transformation function is: $T(Y_i) = \{core - point(C_j \leftrightarrow d(Y_i, core - point(C_j) < d(Y_i, core - point(C_Z) \; \forall \; Z\}$

6. Now all the n transformed portions of each point are joined in the same sequence as portioned in step 2 to form a new k dimension transformed row data which replaces the X in the original dataset. In this way perturbed dataset $D_i{'}$ is formed from original dataset D

7. Compute the privacy level by calculating the difference between the original dataset and transformed dataset using Euclidean distance:

$$\Pr ivacy - Level = \frac{1}{mn} \sqrt{\sum_{i=1}^{m} \sum_{j=1}^{n} \left| x_{ij} - y_{ij} \right|^2} \qquad (3)$$

5.2   Global Perturbation Using G Algorithm

After executing the local obfuscation process, the global perturbation algorithm at the requester side is started. The idea is cluster multidimensional data using fast density clustering algorithm, then perturb each dimension in each cluster in such a way to preserve its range. In order to allow the global perturbation agent to execute G algorithm, we introduce an enhanced mean shift (EMS) algorithm which is tailored algorithm for the global perturbation phase that has advantage over previous algorithm proposed in [17] and it requires low computational complexity in clustering large data sets. we employ Gaussian *KD-tree* [18] clustering to reduce the feature space of the locally obfuscated data.

The G algorithm consists of two steps:

Step 1: Build different density based clusters

1. We build the tree in a top down manner starting from a root cell similar to [18, 19]. Each inner node of the tree S represents a d-dimensional cube cell which stores the dimension S_d along which it cuts, the cut value S_{cut} on that dimension, the bounds

of the node in that dimension S_{min} and S_{max}, and pointers to its children S_{left} and S_{right}. All points in the leaf nodes of the kd tree are then considered as a sample and the kd-$tree$ stores m samples defined as $y_j^*, j = 1, ..., m$ that construct the reduced feature space of the original obfuscated data set.

2. Assign each record x_i to its nearest y_j based on kd-search, then compute a new sample, we called it $y_j^*, j = 1, ..., m$.

3. Generated y_j^* is a feature vector of d-dimensions, that is considered as a more accurate sample of the original obfuscated data set that will be used in the mean shift clustering.

4. The mean shift clustering iteratively performs these two steps:
   — Computation of mean shit vector based on the reduced feature space as following:

$$m(x_j) = \frac{\sum\limits_{y_i^* \in N(y_j^*)} y_i^* g\left(\left\|\frac{x_j - y_i^*}{h}\right\|^2\right)}{\sum\limits_{y_i^* \in N(y_j^*)} g\left(\left\|\frac{x_j - y_i^*}{h}\right\|^2\right)} - x_j, \ j = 1, 2..m \tag{4}$$

Where $g(x) = -k'(x)$ defined when the derivate of function $k(x)$ exists, and $k(x), 0 \le x \le 1$ is called kernel function satisfying: $k(x) = c_{k,d} k\left(\|x\|^2\right) > 0$, $\|x\| \le 1$ and $\int_{-\infty}^{\infty} k(x)dx = 1$

   — Update the current position x_{j+1} as following:

$$m(x_{j+1}) = \frac{\sum\limits_{y_i^* \in N(y_j^*)} y_i^* g\left(\left\|\frac{x_j - y_i^*}{h}\right\|^2\right)}{\sum\limits_{y_i^* \in N(y_j^*)} g\left(\left\|\frac{x_j - y_i^*}{h}\right\|^2\right)}, \ j = 1, 2..m \tag{5}$$

Until reaching the stationary point which is the candidate cluster centre. x_j will coverage to the mode in reduced feature space, finally we get approximate modes of original data defined as $z_x, x = 1,...., k$.

5. Finally, the points which are in the mode are associated with the same cluster. Then we interpolate the computed modes in samples to the original obfuscated data by searching for the nearest mode z_x for each point x_i.

Step 2: Generating random points in each dimension range
For each cluster C, perform the following procedure.

1. Calculate the interquartile range for each dimension A_i.

2. For each element $e_{ij} \in A$, generate a uniform distributed random number r_{ij} in that range and replace e_{ij} with r_{ij}.

6   Experiments

The proposed algorithms are implemented in C++. We used message passing interface (MPI) for a distributed memory implementation of G algorithm to mimic a reliable distributed network of peers. We evaluated the proposed algorithms from two different aspects: privacy achieved and accuracy of results. The experiments presented here were conducted using the Movielens dataset [20]. The dataset contains users' ratings on movies using discrete values between 1 and 5. We follow the experiential scenarios presented in [14] We divide the data set into a training set and testing set. The training set is obfuscated then used as a database for the PCRS. Each rating record in the testing set is divided into a rated items t_u and unrated items r_u. The set $t_{u,i}$ is presented to the PCRS for making predication $p_{u,i}$ for the unrated items $r_{u,i}$ using the same algorithm in [21]. To evaluate the accuracy of generated predications, we used the mean average error (MAE) metric proposed in [22]. The first experiment performed on CBT algorithm to measure the impact of the varying portion size and number of core-points on privacy levels of the transformed ratings. To measure that we kept portion size constant with different number of core-points and then we vary portion size with constant number of core-points. Based on the results shown in figs (3) and (4), we can conclude that the privacy level increases when portion size is increasing. On the other hand, privacy level is reduced with increasing number of core-points as large number of rows used in *TLUT*. Each user in the network can control his privacy by diverging different parameters of *LLA* algorithm. Note that reducing the privacy level means less information loss in the collected ratings presented to PCRS. However this means the transformed ratings are similar to the original ratings, so the attacker can acquire more sensitive information.

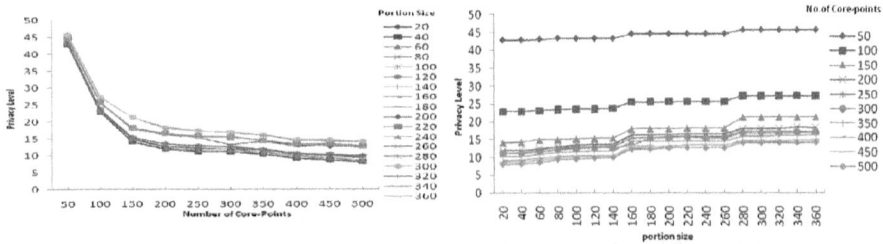

Fig. 3. Privacy level for different no.of core point   **Fig. 4.** Privacy level for different portion size

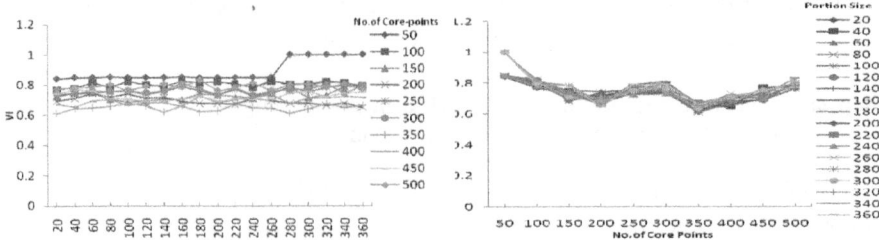

Fig. 5. VI for different portion size   **Fig. 6.** VI for different no.of core points

To measure the distortion of the results, we use variation of information (VI) metric. Fig (5) shows VI for different number of core-points. One can see that at lower values of number of core-points the VI is high but slowly decreases with the increase in the number of core-points. At a certain point it rises to a local maxima then it decreases. Finally it rises again with the increasing number of core-points. We can justify that VI is high with fewer number of core-points as any point can move from one core-point to another. Moreover, with a plenty number of core-points there is a little chance of a point to move from one core-point to another, which causes increasing in VI values. The second experiment is performed on G algorithm to measure the impact of sample size on the accuracy level. We set a specific threshold value (100 users) for the minimum number of responding users for recommendation request. Otherwise the target user uses the PCRS obfuscated ratings database. As shown in fig (7) the increase in sample size leads to higher accuracy in the predications. However at a certain sample size, the accuracy of the predications starts decrement again due to the data loss in the sampling process.

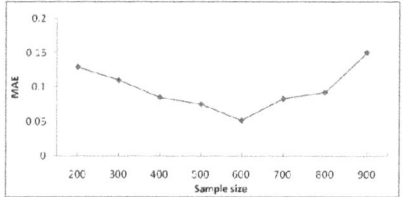
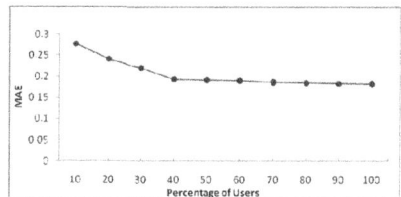

Fig. 7. Relation between sample size and MAE   **Fig. 8.** Relation between Users and MAE

In the third experiment, we want to measure the impact of changing number of users involved in the group formation on the accuracy of the recommendations. We simulate a general case where the number of users was fixed to be 50.000. Then we assign different number of users to a certain recommendation request, and gradually increased the percentage of users who joined the request from 10% to 100% of them. We fixed the parameters for CBT and G algorithms then we measure MAE for the results. As shown in fig (8), the MAE value occurs at approximately 40% of the users are close to the MAE value for all users. Our conclusion is that, for low percentage of users the MAE value is close to the original MAE value for all users. As a result the target user does not need to broadcast the request to the full IPTV network to attain accurate results but he can employ multicast for certain users stored in his peer list to reduce the load in the network traffic. To illustrate the decrement of MAE values for recommendations based on diverse percentages of users groups and the whole users in the network, we calculated and plot fig (9). This verifies our conclusion that MAE approximately converges to the MAE which obtained using the whole users in our case. The final experiment was conducted to measure the impact of using CBT algorithm as a pre-processing step for G algorithm. As presented in fig (10), using CBT increases MAE values for lower percentages of participated users compared to using G algorithm alone. This can be explained, as the distortion effect of CBT algorithm will be clearly visible for a lower percentage of participating users.

However with the augment of percentage of users scale down the error in MAE values. So we can say that using the two stage obfuscation algorithms can increase the recommendation accuracy compared to only one stage based on one algorithm only.

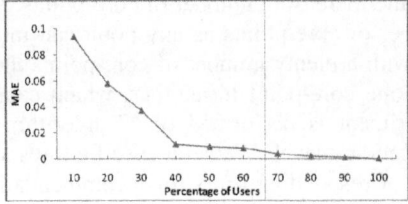   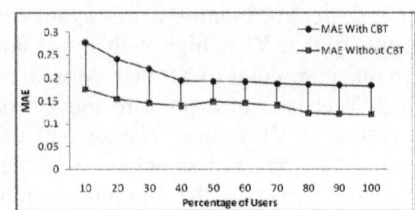

Fig. 9. The decrement of MAE values   **Fig. 10.** The Influence of applying CBT algorithm

The results presented in these experiments show that the resulting dataset from our two stage obfuscation processes are quite similar in the accuracy of the generated recommendation to the original dataset. Our results also clarify that the proposed algorithms preserve the utility of the data to some degree such that to create reliable recommendations the target user does not have to collect profiles from numerous users, only a small percentage from users is need to attain that goal.

7   Conclusion and Future Wok

In this paper, we presented our ongoing work on building an agent based middleware to achieve privacy in recommender services. We gave an overview over the system components and recommendations process. Also we presented the novel algorithms that provide to users complete privacy over his profile privacy using two stage obfuscation processes. We test the performance of the proposed algorithms on real dataset and report the overall accuracy of the recommendations based on different privacy levels. The experiential results shows that preserving users' data privacy for in collaborative filtering recommendation system is possible and the mean average error can be reduced with proper tuning for algorithms parameters and large number of users. We need to perform extensive experiments in other real data set from UCI repository and compare the performance with other techniques, also we need to consider different data partitioning techniques, identify potential threats and add some protocols to ensure the privacy of the data against these threats.

References

1. Hand, S., Varan, D.: Interactive Narratives: Exploring the Links between Empathy, Interactivity and Structure, pp. 11–19 (2008)
2. Cranor, L.F.: 'I didn't buy it for myself' privacy and ecommerce personalization. In: Proceedings of the 2003 ACM Workshop on Privacy in the Electronic Society. ACM, Washington, DC (2003)
3. Dialogue, C.: Cyber Dialogue Survey Data Reveals Lost Revenue for Retailers Due to Widespread Consumer Privacy Concerns. Cyber Dialogue (2001)

 4. Narayanan, A., Shmatikov, V.: Robust De-anonymization of Large Sparse Datasets. In: Proceedings of the 2008 IEEE Symposium on Security and Privacy. IEEE Computer Society (2008)
 5. McSherry, F., Mironov, I.: Differentially private recommender systems: building privacy into the net. In: Proceedings of the 15th ACM SIGKDD International Conference on Knowledge Discovery and Data Mining, pp. 627–636. ACM, Paris (2009)
 6. Esma, A.: Experimental Demonstration of a Hybrid Privacy-Preserving Recommender System. In: Gilles, B., Jose, M.F., Flavien Serge Mani, O., Zbigniew, R. (eds.), 161–170 (2008)
 7. Canny, J.: Collaborative filtering with privacy via factor analysis. In: Proceedings of the 25th Annual International ACM SIGIR Conference on Research and Development in Information Retrieval, pp. 238–245. ACM, Tampere (2002)
 8. Canny, J.: Collaborative Filtering with Privacy. In: Proceedings of the 2002 IEEE Symposium on Security and Privacy, p. 45. IEEE Computer Society (2002)
 9. Polat, H., Du, W.: Privacy-Preserving Collaborative Filtering Using Randomized Perturbation Techniques. In: Proceedings of the Third IEEE International Conference on Data Mining, p. 625. IEEE Computer Society (2003)
10. Polat, H., Du, W.: SVD-based collaborative filtering with privacy. In: Proceedings of the 2005 ACM Symposium on Applied Computing, pp. 791–795. ACM, Santa Fe (2005)
11. Huang, Z., Du, W., Chen, B.: Deriving private information from randomized data. In: Proceedings of the 2005 ACM SIGMOD International Conference on Management of Data, pp. 37–48. ACM, Baltimore (2005)
12. Kargupta, H., Datta, S., Wang, Q., Sivakumar, K.: On the Privacy Preserving Properties of Random Data Perturbation Techniques. In: Proceedings of the Third IEEE International Conference on Data Mining, p. 99. IEEE Computer Society (2003)
13. Parameswaran, R., Blough, D.M.: Privacy preserving data obfuscation for inherently clustered data. Int. J. Inf. Comput. Secur. 2, 4–26 (2008)
14. Elmisery, A., Botvich, D.: Private Recommendation Service For IPTV System. In: 12th IFIP/IEEE International Symposium on Integrated Network Management. IEEE, Dublin (2011)
15. Blaze, M., Schneier, B.: The MacGuffin block cipher algorithm, pp. 97–110 (1995)
16. Elmisery, A.M., Huaiguo, F.: Privacy Preserving Distributed Learning Clustering Of HealthCare Data Using Cryptography Protocols. In: 34th IEEE Annual International Computer Software and Applications Workshops, Seoul, South Korea (2010)
17. Fukunaga, K., Hostetler, L.: The estimation of the gradient of a density function, with applications in pattern recognition. IEEE Transactions on Information Theory 21 (2003)
18. Xu, K., Li, Y., Ju, T., Hu, S.-M., Liu, T.-Q.: Efficient affinity-based edit propagation using K-D tree. In: ACM SIGGRAPH Asia 2009 Papers, pp. 1–6. ACM, Yokohama (2009)
19. Adams, A., Gelfand, N., Dolson, J., Levoy, M.: Gaussian KD-trees for fast high-dimensional filtering. ACM Trans. Graph. 28, 1–12 (2009)
20. Lam, S., Herlocker, J.: MovieLens Data Sets. Department of Computer Science and Engineering at the University of Minnesota (2006)
21. Herlocker, J.L., Konstan, J.A., Borchers, A., Riedl, J.: An algorithmic framework for performing collaborative filtering. In: Proceedings of the 22nd Annual International ACM SIGIR Conference on Research and Development in Information Retrieval, pp. 230–237. ACM, Berkeley (1999)
22. Herlocker, J.L., Konstan, J.A., Terveen, L.G., Riedl, J.T.: Evaluating collaborative filtering recommender systems. ACM Trans. Inf. Syst. 22, 5–53 (2004)

Privacy Enhanced Device Access

Geir M. Køien

University of Agder
Department of ICT, Grimstad, Norway
geir.koien@uia.no

Abstract. In this paper we present the case for a device authentication protocol that authenticates a device/service class rather than an individual device. The devices in question are providing services available to the public. The proposed protocol is an online protocol and it uses a pseudo-random temporary identity scheme to provide user privacy.

Keywords: Authentication, Key derivation, Device access, Internet-of-Things, Security protocol, Privacy, Public access.

1 Introduction

1.1 Background

In this paper we present the case for a lightweight device access protocol that authenticates a device/service class rather than an individual device. The background for this is the observation that, for a human user (through his/her proxy device), access is often towards a service rather than any specific device.

A typical public Internet-of-Things (IoT) device is often a nameless device as seen from the human users perspective. An IoT device typically consist of an embedded computing platform with various sensors and actuators, and it has IP-connectivity (wired/wireless). There is little incentive for the user to know the identity (serial number or similar) of the device per se. Except of course, that this is often a prerequisite for connectivity. We assume that the IoT devices are accessible via a "broadcast" type of channel. Thus, the user cannot ascertain a device merely by means of physical connectivity (i.e. by fixed line cabling).

The devices are managed by an administrative entity and it will issue device access credentials to the user. Depending on the type of service requested it should be possible to access multiple devices during the same service period.

1.2 Use-Cases

We propose three example services where our device/sevice-access protocol could be deployed. Example cases:

- **C1: Parking Permit**
  A parking lot entrance is controlled by an IoT device. The permit may apply to any parking facility operated by the granting authority or for a a single location. The permit may be for a single access or for a pre-defined period.

R. Prasad et al. (Eds.): MOBISEC 2011, LNICST 94, pp. 76–87, 2012.

- **C2: Auto Wash Machine**
  The auto wash machine is controlled by an IoT device. Washing rights can used for any auto wash machine operated by the same management. The washing right may be for one or more washes and for different types of washing.
- **C3: Wi-Fi Login Services**
  The user may buy access right for a period (D days, M months, ...) with the Wi-Fi operator. The access right could apply to all Wi-Fi access points within the operators network.

1.3   Outlining the Problem

We here list some aspects that must be considered for our device access protocol.

- *Flexible service authorization during access*
  Flexibility is needed when defining what an "access right" amounts to. Time period or a defined no. of access events? Specific device or any device? It may be beneficial to let the device operator handle the service authorization part.
- *Online vs. offline*
  Is the device operator to be online or offline during device access? There are pros and cons to both offline and online models. Requirements for authentication, authorization and accounting point towards online models, while communications cost and possibly user privacy will benefit from an offline model.
- *No Group Keys*
  As stated we want access to be for a service type rather than for any specific device. However, we do not want a scheme with a symmetric group key for a all devices with the given service type, since we consider group keys to be a security and privacy liability. We argue here that the IoT devices in question will be widely distributed and that while physical protection will be in place, it will nevertheless be limited. Then it is important that the compromise of one device does not unduly lead to compromise of other devices.
- *User privacy*
  We insist that the user be given credible privacy protection with respect to identity privacy and location privacy. This requirement is important not only toward external parties, but is also of importance internally. In particular, we want to limit the IoT devices ability to compromise the privacy of the users.
- *A Preference for Simplicity*
  The device/service access protocol should be as simple and lightweight as possible with respect to computational and communication needs.

2   Reference Architecture and Assumptions

2.1   Principal Entities

We will have three different types of principal entities:

- OPR : An administrative entity which owns and/or operates the devices.
- USR : The user is represented by a proxy device (possibly a a smart phone).
- DEV : The generic IoT device, which will provide/facilitate services.

As is customary we assume that we have honest principals, that will faithfully conduct their business as intended. However, we will also have a defined intruder, which we denote DYI, in our system and this intruder is free to masquerade as a principal. More on our intruder in subsection 3.1.

2.2   Generic Architecture

Figure 1 depicts the architecture, in which the user (USR) has gained access to a device (DEV 3). We chose to model a one-to-one correspondence between a service and the associated provider. Obviously, multiple provider may offer the same basic service and one provider may offer multiple distinct services. If one models the different consumers, the different services and the different providers as distinguishable, over the interfaces, then a richer model can be defined, but the essential properties will not differ in the logical layout of the models. For sake of simplicity we only depict a simplified model in figure 1.

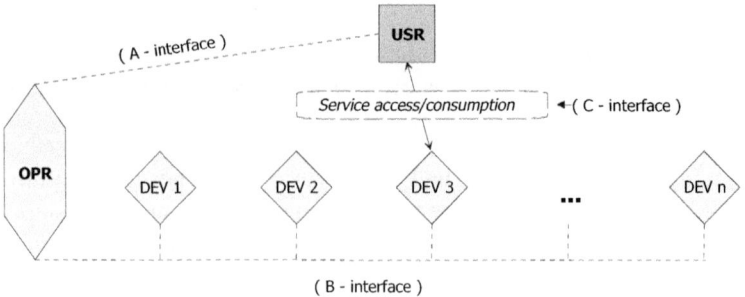

Fig. 1. Simplified Architecture

2.3   Interfaces and Channels

Between the principal entities we define the following interfaces:

- A-interface: Between OPR and USR
- B-Interface: Between OPR and DEV
- C-interface: Between USR and DEV

Each interface is associated with a logical channel. For sake of simplicity, the channel is named after the interface.

The A-interface/A-channel is used for service agreement. This includes all agreements between the user and the operator, and will include exchange of security credentials to facilitate service access between the user and the devices. We shall not assume that the A-interface is operational during service access.

The B-interface/B-channel is for communication between the device and the operator. This includes device initialization and it may include session authorization during user device access.

The C-interface/C-channel is defined between the user and the devices. The associated C-channel is established during service access setup and will be available during service consumption.

2.4   Communication Security Assumptions

The A-channel is assumed to be an authenticated and fully secured channel. Security for the A-channel is assumed to be pre-arranged and the actual setup and agreement of security credentials and key material for the A-channel is considered outside the scope of this paper.

The B-channel is assumed to be an authenticated and fully secured channel. The operators own/manages the devices so it seems reasonable to assume that security credentials and key material is simply distributed by the operator to the devices during device deployment.

The C-channel will need to be mutually authenticated, with respect to service consumption rights, and it will need to be protected. The protection in question should certainly include data integrity protection. Digital Rights Management and user privacy may dictate data confidentiality too, and we claim that it in general is best to assume data confidentiality as a requirement.

2.5   Computational Performance

We do not expect the computational requirements, with respect to provisioning access, to be prohibitive if one mainly designs the security and privacy part of the protocol around symmetric-key cryptographic primitives.

Neither do we in general anticipate that careful and discriminate use of normal public-key primitives are problematic. However, that assumption may be at odds with some devices, like passive RFID devices or devices with limited battery power etc, and for those cases one may find the requirements to be computationally exhausting for frequent access. Thus there is a case for making the access protocol as lightweight as possible with respect to computational requirements.

2.6   Assumed Trust Relationships

We have the following trust relationships:

- **Trust** OPR $\rightleftarrows$ DEV
  The operator has full security jurisdiction over the devices. Devices are plentiful and may become corrupted or may otherwise fail. The operator therefore

cannot afford to trust the device too much. Trust has to be contained such that compromise of one device would not unduly lead to compromise of other devices. The devices must trust the operator fully.

- **Trust OPR→USR**
  This relationship is governed and limited by contractual agreement.
- **Trust USR→OPR**
  The user must trust the operator with respect to service purchase and consumption. Privacy: The user does not fully trust the operator with information regarding service consumption (time/location etc), but must trust the operator with payment information and possibly with identity information.
- **Trust USR ⇌ DEV**
  There is no a priori security trust between these entities. All trust must be through operator mediation. The established trust depends on trust transitivity. Transitive trust is indirect and we assume it to be weaker than direct trust. The trust level between the user and device is affected by whether the operator is online or not. For the offline case, one cannot know whether the operator still authorizes the access. The privacy trust that a user may have in a device is strictly limited to the needs for service access.

3   Intruder Model and Intruder Mitigation

3.1   Intruder Model

The classical intruder model is the well known Dolev-Yao (DY) intruder model [1]. In this model the intruder can intercept, modify, delete and inject messages at will. The intruder will have complete history information of all previous message exchanges and it can (and will) use this to the full extent to attack the system. It may also masquerade as a legitimate principal. The DY intruder is also a privacy intruder and it will exploit leakage of privacy sensitive data to the full extent.

The DY intruder is a very powerful adversary, but it has its limitations. For instance, it cannot physically compromise a principal and it cannot actually break cryptographic primitives (but it can exploit improper use). It may appear somewhat contrary that it cannot break "weak" cryptographic primitives, but this is nevertheless the standard assumption.

In the general case for an IoT environment one must expect devices to break down or otherwise be unable to maintain their physical integrity. Thus, with statistical validity, one must expect devices to be compromised and secrets to be exposed. The intruder will, by definition, exploit these shortcoming and will thus learn whatever secrets the IoT device contains, including privacy sensitive user data. The exploitation will potentially continue for the lifetime of the device.

However, while we advocate using the DYI model for analysis we still want to be realistic with respect to real-life intruders. These intruder will rarely be capable of using all available knowledge, and so pose less of a danger to the system. On the other hand, real-world intruder are capable of breaking "weak" crypto primitives and crypto system with short keys etc.

3.2 Compromise Containment

The compromise of one or more devices should not give the intruder an advantage in compromising other principal entities, whether devices, operators or users. Therefore, we require that secrets and sensitive data (key material, identities etc) stored at the devices should not unduely permit the DYI to compromise other principals.

With respect to cryptography we require the key sizes etc to be "sufficiently long", and we here advise adhering to the ECRYPT II recommendations [2].

4 Security Requirements and Privacy Aspects

4.1 Security Requirements

The user must verify that the device is authorized to provide services (by the operator). Correspondingly, the device must be assured that the user has obtained the rights to access the indicated service. The requirements are fairly easy to satisfy if the operator is online during the device access. If the B-interface is unavailable during device access, then the access protocol must provide assurance that the access once was sanctioned by the operator.

A suitable set of session keys must be agreed for the device access and service access procedures.

4.2 The Need for Privacy

The user identity, whether associated or emergent, should not be disclosed to any unauthorized parties. Also, the user location should not unduely be disclosed and tracking of the user should not be permitted. Data privacy, in conjunction with service payload and with identifying the consumed service must also be provided.

We note that the devices are not fully trusted with respect to user privacy. That is, the devices (DEV) are partially an unauthorized party. The operator (OPR) must be trusted by the user USR, but we still want the operator knowledge of privacy sensitive data to be limited to a minimum.

5 Online Model vs. Offline Model

5.1 Online Model

For an online case one can simplify the the protocol structure significantly and provide an assurance level that simply is not attainable with offline protocols [7]. The operator provides assurance and can forward key material to the device upon request, and thus the user and the device need not have complete credentials to start off with. This makes it is feasible to carry out all operations with symmetric-key primitives. This is an advantage when designing a computationally lightweight protocol solution.

5.2 Offline Model

To achieve satisfactory security for the offline case is not easy with the requirements that the device should not need to know the identity of the user.

To achieve the given goals, and to avoid using group keys, it is necessary to use asymmetric crypto primitives. To keep things simple we suggest that the operator issue short-lived digital certificates to the user. These digital certificates could be service-specific. The device verifies that the user had a legitimate "access certificate" and grants access accordingly. Since there already exists several suitable public-key/digital certificate protocols we suggest that one then deploys such protocol instead of inventing a new protocol. We suggest using TLS (TLS v1.2, RFC 5246 [3]). Certificate revocation must be checked (RFC 3280 [4]), but in an offline model this information may be obsolete.

6 Outline of Online Solution

The following section outlines the "Privacy Enhanced Lightweight Device Authentication" (PELDA) protocol.

6.1 Protocol Outline

Figure 2 depicts the main PELDA protocol.

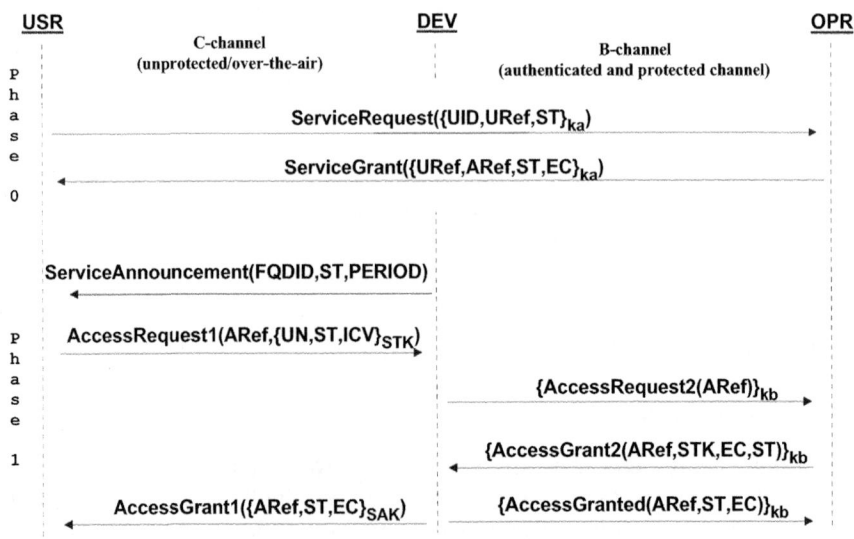

Fig. 2. Outline of the PELDA protocol

6.2   Information Elements

In accordance with ECRYPT II [2] recommendations we advocate all keys to be at least 128 bit long. We also mandate that the identities, references and (pseudo-random) nonces be 128 bit long. We will otherwise remain agnostic with respect to the concrete encoding of the identifers.

```
UID   : Globally Unique Permanent privacy-sensitive user identity;
DID   : Globally Unique Permanent public device identity;
SID   : Permanent public server identity;
FQDID : SID||DID; Fully Qualified Device Identity;
URef  : User Reference; Also URef';
ARef  : Access Reference;
UN    : User Nonce;
ST    : "Service Type" identifier;
EC    : "Expiry Condition" identifier;
ICV   : Integrity Check Value;
STK   : Service Type Key;
SAK   : Service Access Key;
TAKx  : Temporary Access Key x; x={i|c}, (integrity|confidentiality)
```

Fig. 3. PELDA phase 1 - Information Elements

6.3   Key Derivation

During PELDA execution three different symmetric keys are derived. These are the "Service Type Key (STK)", the "Service Access Key (SAK)" and the "Temporary Access Key $(TAKx)$". The STK and SAK are key deriving keys.

The TAK is used for protection of service content, and it consists of separate integrity $(TAKi)$ and confidentiality $(TAKc)$ keys. One could also distinguish between uplink and downlink keys, but we have chosen not to do so here.

The Service Type Key (STK) is a key derived for a specific user identified by the "user reference" $(URef)$. The key derivation transform uses a symmetric-key encryption function, E, and the bitwise exclusive-or function. The $kdf1$ function has the property that even a party that possess the "input key" and the output key will not be able to deduce the input parameter. The "input key" in question is not an ordinary encryption key, but the service type identifier (ST). The input parameter $URef'$ is a MAC-modified pseudo-random user reference. The $URef$ is required to be confidential to all but the user (USR) and the operator (OPR). The encrypt-and-xor transform it is not new. It was also used in the MILENAGE authentication algorithm set used in cellular systems (UMTS and LTE) [6].

$$kdf1_{ST}(URef') \rightarrow STK \equiv (E_{ST}(URef') \oplus URef') \rightarrow STK \qquad (1)$$

The Service Access Key (SAK) is specific to one device. The key derivation function takes STK as the input key. The input parameters, all non-secret, includes the device identity (DID), the service type (ST), user generated pseudo-random nonce UN and the access reference ($ARef$). The $ARef$ is generated by the operator and must be guaranteed to be unique with respect to the request context (given in the ServiceRequest message).

$$kdf2_{STK}(FQDID, ARef, ST, UN) \rightarrow SAK \tag{2}$$

The Temporary Access Key (TAK) is specific to a session or period. The key derivation function takes the SAK as the input key. The only input parameters is the $PERIOD$ identifier (which was broadcast by the device). Whenever the $PERIOD$ indication in the ServiceAnnouncement changes the TAK keys must be re-computed. The expiry condition (EC) may not necessarily coincide with the $PERIOD$ announcement, and that expiry of EC may lead to expiry of the whole service context irrespective of $PERIOD$ expiry.

$$kdf3_{SAK}(PERIOD) \rightarrow TAKi \| TAKc \tag{3}$$

6.4   PELDA Protocol Description

We now present the PELDA protocol in an augmented Alice-Bob notation. We assume the presence of a pseudo-random number function (prf), suitable (block cipher) symmetric-key primitives and a MAC function. We also assume that keys for the A-channel (ka) are agreed prior to PELDA execution and that keys for the B-channel (kb) are available during PELDA phase 1 execution.

PELDA Phase 0 - Access Agreement

```
0. Preparations (pre-computation possible)
   · USR: prf(·) → URef

1. USR→OPR: ServiceRequest({UID, URef, ST,}ka)
   · OPR: Decrypt the ServiceRequest message and prepare the response.
   · OPR: prf(·) → ARef

2. OPR→USR: ServiceGrant({URef, ARef, ST, EC}ka)
```

PELDA Phase 1 - Initial Registration

```
0. DEV→all: ServiceAnnouncement(FQDID,ST,PERIOD)
   · USR: prf(·) → UN
   · USR: MACka(URef, FQDID) → URef'
   · USR: kdf1ST(URef') → STK
   · USR: kdf2STK(FQDID, ARef, ST, UN) → SAK
   · USR: MACSTK(FQDID, PERIOD, ARef, ST, EC) → ICV
```

1. USR→DEV: AccessRequest1(ARef,$\{UN, ST, ICV\}_{STK}$)
2. DEV→OPR: AccessRequest2($\{ARef\}_{kb}$)
   - OPR: $MAC_{ka}(URef, FQDID) \rightarrow URef'$
   - OPR: $kdf1_{ST}(URef') \rightarrow STK$

3. OPR→DEV: AccessGrant2($\{ARef, STK, EC, ST\}_{kb}$)
   - DEV: Decrypt $\{UN, ST, ICV\}_{STK}$ from AccessRequest1
   - DEV: Verify: $MAC_{STK}(FQDID, PERIOD, ARef, ST, EC) = ICV$
   - DEV: $kdf2_{STK}(FQDID, ARef, ST, UN) \rightarrow SAK$

4. DEV→USR: AccessGrant1($\{ARef, ST, EC\}_{SAK}$)
   DEV→OPR: AccessGranted($\{ARef, ST, EC\}_{kb}$)
   - USR,DEV: $kdf3_{SAK}(PERIOD) \rightarrow TAKi\|TAKc$

PELDA Phase 2 - Rekeying

Re-keying of $TAKx$ takes place when the $PERIOD$ identifier in the broadcast message ServiceAnnouncement changes.

7   Protocol Analysis

7.1   Complexity and Efficiency Aspects

The computational cost of the PELDA protocol is modest and only symmetric-key primitives are used. We foresee no problem in this area and see no need for a more detailed analysis of this aspect.

We note that with a cipher block size of 128 bits, all messages will conformably fit within 4-5 blocks. This makes the communications complexity quite modest and we see no need for a more detailed analysis of this aspect either.

With respect to round-trip delays, a full round-trip from the user, via the device and to the operator is required. The delay should not be prohibitive, and since we have required the operator to be online a full round-trip cannot be avoided. Observe that AccessGrant1 and AccessGranted are sent in parallel and thus do not induce any additional delay. With this in mind, we conclude that the PELDA phase 1 protocol is indeed a lightweight protocol.

7.2   Brief Security Analysis

PELDA Phase 0: We have that the A-channel is authenticated and protected. Thus, we postulate that security is maintained for the A-channel.

PELDA Phase 1: The user initially generates the $URef$ and UN elements, which are pseudo-random and which are unique and unpredictable. The STK is derived by the user and is known to user to be a fresh secret key. The $ARef$ is known to the user to be associated with the fresh $URef$ (phase 0). The user can thus assume that the $ARef$ is fresh and unique.

In `AccessRequest1` the user sends $ARef$ to the device. The device forwards $ARef$ to the server over authenticated and fully protected the B-channel. The device fully trusts the operator. Thus, when the device receives the `AccessGrant2` message it has assurance that the $ARef$ is valid and that it is associated with STK, EC and ST. By the ICV the device also has assurance that access attempt is indeed for it, for the $PERIOD$ (timeliness) and for and for the user ($ARef$). Subsequent to `AccessGrant1` the user has assurance that the device was recognized by the server through the use of SAK. (SAK can only be derived by a party which knows STK and $ARef$). The user already has assurance of STK and $ARef$, and it therefore accepts the device as being valid.

The server does not get explicit assurances of the user during PELDA phase 1, but relies on the device to ascertain the user ($Aref$). However, the server has instructed the device (in `AccessGrant2`) only to provide services as agreed for $ARef$ (encoded in ST, EC), and so it has covered its needs.

7.3   Brief Privacy Analysis

Our main privacy requirements are that the device should not be allowed to know the user identity. Since the device is never actually given the user identity, neither the UID or the URef, we may conclude that the requirement is trivially fulfilled. However, improper use of the $ARef$ could lead to an intruder constructing an emergent identity for the user. Thus, one must assure that the $ARef$ is not exposed (`AccessRequest1`) too many times.

8   Summary and Concluding Remarks

8.1   Summary

We have presented the privacy enhanced lightweight device authentication (PELDA) protocol. The goal of the protocol was to facilitate access to publicly operated IoT-based services such that the user may concentrate of service access rather than on device access.

Privacy is a concern and in particular we have the concern that inexpensive and widely distributed IoT devices may not provide the best protection of privacy sensitive data. The PELDA protocol was designed with this in mind and it will not unduely store privacy sensitive data at the devices. Furthermore, since the PELDA protocol uses a disposable "access references" in place of a permanent identity, there is very little privacy information that the device can leak/divulge. Corollary, subscriber privacy will suffer if $ARef$ is re-used for a prolonged period. The drawback to the PELDA scheme is that it requires the operator to be online, with respect to the device, during the initial device access. This is an unavoidable consequence of the PELDA protocol requirements, but we do not see this as a very limiting restriction in a future with almost pervasive internet connectivity. The bare-bones PELDA protocol is a simple protocol with few roundtrips, a relatively small payload and modest crypto-performance requirements.

8.2 Further Work

We intend to implement variants of the PELDA protocol with an aim to investigate how these protocols cover a wider set of use-cases. We also intend to develop formal models and carry out formal verification of selected aspects. We intend to use the AVISPA (`www.avispa-project.org`) and/or the AVANTSSAR (`www.avantssar.eu`) tools. Here of course we would in particular investigate privacy properties (the formal security modeling tools tend to cater well for entity authentication etc already). Still, we tend to agree with Gollmann [8] in his reluctance to trust formal verification to prove any protocol correct (Gollmann even refers to proofs as a non-goal of formal verification).

Finally, we add that security, privacy and performance comparisons with other alternatives should be conducted. Amongst the alternatives are the "Identity and Access Services" platform from the Kantara Initiative (formerly Liberty Alliance) and solutions based on the US federal initiative "Open Identity Solutions for Open Government" (More information at `http://www.idmanagement.gov.`).

8.3 Concluding Remarks

We have presented the PELDA protocol and we currently believe that it is an adequate protocol for device access. There remains aspects of the protocol that need deeper analysis, but initial investigations seems to indicate that it is secure and that it provides credible user privacy. Furthermore, the simplicity of the protocol indicates that it will perform well and it should scale well too.

References

1. Dolev, D., Yao, A.: On the security of public key protocols. IEEE Transactions on Information Theory 29(2), 198–208 (1983)
2. Smart, N. (ed.): ECRYPT II Yearly Report on Algorithms and Keysizes (2009-2010). Rev.1, 30. March 2010, ECRYPT II (ICT-2007-216676), European Network of Excellence in Cryptology II (2010)
3. Dierks, T., Rescorla, E.: RFC 5246: The Transport Layer Security (TLS) Protocol Version 1.2, IETF, 08-2008
4. Housley, R., Polk, W., Ford, W., Solo, D.: RFC 3280: Internet X.509 Public Key Infrastructure Certificate and Certificate Revocation List (CRL) Profile, IETF (April 2002)
5. Køien, G.M.: Entity Authentication and Personal Privacy in Future Cellular Systems. River Publishers, Aalborg (2010)
6. 3GPP, TS 35.205: 3G Security; Specification of the MILENAGE Algorithm Set: An example algorithm set for the 3GPP authentication and key generation functions f1, f1*, f2, f3, f4, f5 and f5*; Document 1: General (Release 9), Sophia Antipolis, France (December 2009)
7. Boyd, C., Mathuria, A.: Protocols for Authentication and Key Establishment. Springer (1998)
8. Gollmann, D.: Analysing Security Protocols. In: Abdallah, A.E., Ryan, P.Y.A., Schneider, S. (eds.) FASec 2002. LNCS, vol. 2629, pp. 71–80. Springer, Heidelberg (2003)

Energy Efficiency Measurements
of Mobile Virtualization Systems

Marius Marcu and Dacian Tudor

"Politehnica" University of Timisoara
2 V. Parvan Blv, 300223 Timisora, Romania
{mmarcu,dacian}@cs.upt.ro

Abstract. The energy efficiency has become an important aspect in data centers and large server systems, including the ones used in infrastructure for mobile applications service providers. Virtualization is one of the main research directions for both large scale data centers and applications servers. Furthermore, virtualization is also popular on desktop systems and is now considered in embedded systems. The next step will be to use virtualization on battery powered systems or mobile devices, where power consumption is an important aspect. This paper explores how virtualization influences the power consumption of both physical systems and virtual systems and which is the most efficient way to implement virtualized applications. The paper proposes a test bench and a set of test cases which can be further used to evaluate and compare different virtualization solutions together with several power management mechanisms using specific energy efficiency metrics.

Keywords: consumption, energy efficiency, virtualization, virtual machine.

1   Introduction

We have been witnessing the development of enterprise servers and data centers to support cloud computing for the last few years. The design of data and computing centers implies a tradeoff between performance and power consumption. The main requirement for these solutions is to provide the agreed level of services while trying to minimize the service provisioning costs [1]. Power consumption is a critical parameter in modern datacenter and enterprise environments, since it directly impacts both the deployment costs (peak power delivery capacity) and operational costs (power supply, cooling) [2]. One solution for power consumption reduction is to consolidate multiple servers running in different virtual machines (VMs) on a single physical machine (PM) which increases the overall utilization and efficiency of the equipment across the whole deployment. [3]

On the other hand we assist to an increasing development of mobile applications and services intended for various types of mobile devices. This trend will influence other areas, including the cloud computing solutions. Cloud computing support for mobile applications is identified as a new direction of research and development. Although several research works have been conducted in the field of cloud computing

R. Prasad et al. (Eds.): MOBISEC 2011, LNICST 94, pp. 88–100, 2012.

for mobile technologies, this field is vastly unexplored [5]. Cloud computing for mobile applications and service, called Mobile Cloud Computing is a well-accepted concept that aims at using cloud computing techniques for storage and processing of data on mobile devices, thereby reducing their limitations [5]. Several characteristics make mobile applications special related to other types of applications executed in the cloud: large number of users, devices with small amount of resources, forward complex operations to run on the cloud, different usage pattern, security services, etc. Therefore we can say that mobile devices will be prepared in near future to implement specific virtualization solutions.

Our main research goal is to investigate the energy efficiency of virtualization solutions in battery powered computing systems. The paper proposes a test bench and a set of test cases which can be further used to evaluate and compare different virtualization solutions together with several power management mechanisms using specific energy efficiency metrics. In our current tests we investigate energy efficiency of several algorithms or benchmarks (memory, IO and CPU) and different user applications. We execute the proposed tests on a dual-core laptop with L4 Linux paravirtualization solution.

Section 2 of this paper contains a brief look at energy efficiency of virtual platforms and specific power management mechanisms available for VMs. In section 3 we define the evaluation methodology used in our tests to show the power consumption and energy efficiency of virtual systems. In section 4 we present the results we obtained for power consumption and energy efficiency proposed test cases.

2 Power Management of Virtualized Solutions

A cornerstone in the energetically evaluation for virtualized systems is the measurement procedure and context for both physical systems and virtual machines. Power consumption of physical servers is an important metric used when evaluating different virtualized solutions implemented on top of these servers. The power consumption issue of computing systems is in general a very complex one because every physical component in the physical system has its own power consumption profile depending especially on its execution workload. In virtualized environments the power consumption modeling problem is much more complex because software applications are running on VMs and they do not access directly the physical components. The host operating system has to provide access to physical components and share these components for different VMs and their applications. The nature of workload executed in each VM determines the power profile and performance of the VM, and hence its energy consumption [2]. The complexity of measuring energy efficiency for virtualized systems is increasing with the number of elements that should be addressed (e.g. number of VMs, OS, PM, power management mechanisms activated, software applications running on VMs).

The author of [1] proposed and performed a set of virtualization performance tests for three types of Intel multi-core based servers in order to estimate whether their virtualization can deliver significant benefits in data centers over non-virtualized servers. During the performance tests overall power consumption was measured and power consumption per workload was computed in order to determine the costs of providing the requested level of performance. Virtualization enables one to consolidate multiple workloads onto each server, increasing utilization and reducing power consumption per workload [1]. A CPU intensive complex database application was used as testing workload, and they progressively increased the number of virtualized workloads. In our approach we use three types of simple operations as workload in order to address the main components of the system: CPU, memory and I/O.

Another important element in energy efficiency evaluation for virtualized systems is related to the power management mechanisms and their implementations. The authors of [2] presented a multi-tier software solution for energy efficiency computing in virtualized environments based on the characteristics of the workloads co-located on the same PM. The paper shows that co-location of VMs with heterogeneous characteristics on same PM is beneficial for overall performance and energy efficiency. In [7] the authors investigate the design, implementation, and evaluation of a power-aware application placement controller in the context of an environment with heterogeneous virtualized server clusters. Their solution dispatches applications to different VM or PM taking in account performance requirements, migration costs and power consumption. The tests and experiments were executed based on the traces obtained from server farm of a large data center.

In [4] specific work related to power management of virtualized OS is presented. The authors tried to map virtual ACPI power states of VM components (e.g. CPU P-states, OS S-states and devices D-states) to real power states of the PM in order to increase the efficiency of overall power management mechanism. Nathuji and Schwan explored in their work how to integrate power management mechanisms between VMs and host PM while keeping isolation between them [6]. They proposed and implemented a software solution called VirtualPower which extends the VM power states and assign specific power policies to these states. Their main challenge is again to map VM power states to real power states of the PM.

The authors of [8] focused their research work to power management of I/O disk operations in virtualized environments. This paper presents three proposed improvements to address the disk's device drivers' power states mapping between VM and PM, based on the statistics of I/O activities between PM and VM. Their solutions are based on different combinations between buffering mechanism in the PM that buffers writes from the VMs and early flush mechanism that flushes the dirty pages from the guest OS buffer caches prior to putting the disk to sleep.

A major challenge in computer systems is the coexistence of real-time and non-real-time applications on the same machine. The authors of [9] describe the microkernel architecture of L4 and which provides both virtualization and real-time support. On a real-time capable microkernel, all applications are temporally isolated and can execute with real-time guarantees even they are virtualized. L4 Linux is a paravirtualization solution which requires changes in the guest operating systems in order to run in user space of the CPU. The changes are required in platform-specific code of Linux but all other code and device drivers are unchanged. L4 Linux was ported on IA-32 and ARM processors architectures; therefore it may be used in the near future on next multi-core mobile devices.

3   Energy Efficiency Virtualization Evaluation Methodology

In this section we describe the evaluation methodology we propose to estimate energy efficiency of virtualization solution implemented on different physical computing systems. The proposed methodology describes two aspects: evaluation test bench and evaluation test cases. First, the evaluation test bench contains the testing setup used to collect power consumption and workload performance data, to control the workload execution sequence and to provide support for energy efficiency computation and analysis. Second, the evaluation test cases specify the workload applications and configurations scenarios used to emphasis the effect of virtualization over the physical guest system power consumption.

The proposed evaluation methodology describes a standard way to evaluate power consumption and energy efficiency of VMs running on different common hardware. This methodology can be further used to compare energy data for various combinations of physical hardware, operating systems and virtualization solutions.

3.1   Evaluation Test Bench

The proposed test bench is used for power consumption and energy efficiency evaluation of L4 Linux microkernel based virtualization solutions. Test bench description contains the physical hardware systems where the VMs are running, the operating systems installed on these physical and VMs and the power measurement devices (Fig. 1).

Experimental evaluation of system virtualization power consumption uses standard multicore desktop and laptop hardware. Hardware test configuration contains two different machines, both running Ubuntu Linux and L4 Linux:

(1) HP Compaq dc7800 desktop with Intel Core2 Quad 2.4 GHz CPU, 4 GB of memory and 400 GB hard disk, and

(2) HP EliteBook 8530w laptop with Intel Core2 Duo 2.53 GHz processor, 2 GB of memory and 140 GB hard disk.

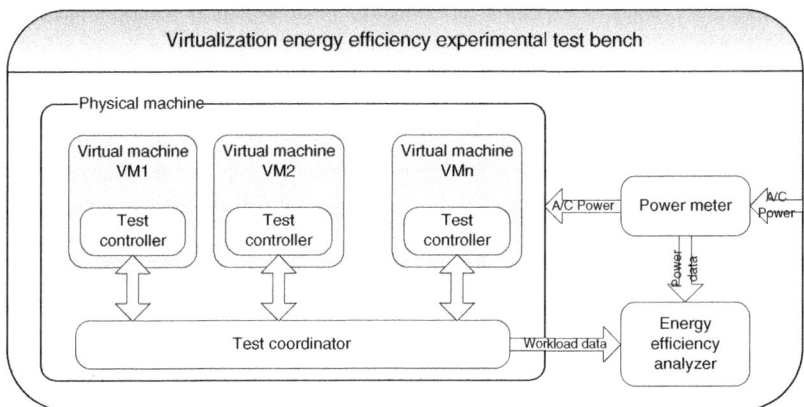

Fig. 1. Overall evaluation setup architecture

Physical machines under tests have both host operating system and virtualization system installed in order to test energy efficiency of workload operation in both PM and VM. Power measurements are considered in both absolute values and relative values to the idle state power consumption. Power measurements for both desktop and laptop systems were obtained using Watts up? series power meters. Power consumption of the entire physical hardware is measured on A/C power lines. Power measurements were obtained with a sampling rate of one per second and were saved in log files for further offline analysis. The Watts up power meter logged power values locally thereby avoiding undesirable measurements effects on the machine under test.

3.2   Evaluation Test Cases

The overall power consumption of the whole device is composed of power consumption of every device's component and the software applications running on that device. Based on this assumption we consider that software applications have a certain level of power consumption. In our proposed tests we try to estimate power consumption of physical system when the virtualization solution is running a number of workloads. Both virtualization solution and workload tests are software applications which have to be estimated from the energy perspective. Power consumption of software applications are hard to estimate or compute due to their uncertainty and interference with other running applications. Therefore we need a standard set of test cases which can be run in order to obtain an estimation of virtualization power consumption and energy efficiency with good accuracy and low measurements dispersion.

During each phase of the test case the power measures are achieved with a rate of one per second and the minimum and average power values are accounted. Every interval in the test profile lasts for a certain amount of time (e.g. 5 to 10 minutes) when no other applications, user inputs or communications are allowed. Also during the test, power management transitions are prevented to occur in order to measure exactly the workload under test. Every test is executed directly on the PM and on the virtual machine. The test execution is coordinated by a central component running on the PM (Fig. 1). This test coordinator establish a connection with all VMs in the system and start the test sequence within every VM according with the test case selected pattern.

When the test workload is running other system's parameters are read and saved in log files. CPU parameters like CPU usage, cores usage and cores temperatures are some parameters we also logged during the tests. We tried also to extract specific performance information for the workload phase in order to compute power efficiency for every executed test. The results obtained when tests are ran show how power consumption of PM varies during workload execution relative to the idle state power consumption. The workload could be executed on PM or VM. When the same test is ran more than once in the same conditions (e.g. PM or VM), the same power profile was obtained.

The data logged during test execution are analyzed offline based on power signature plotted from these data and further power levels and power efficiency values are computed for each test.

Idle State Physical System Power Consumption

The first test case we run on every system under test was introduced to estimate power consumption of the physical system when running in idle state when no power management profile is selected on the host operating system. We consider that the system idle power consumption is important in because the workloads' power consumption introduced in the next test cases will be estimated compared with this initial value. We name this test case IDLE_PHY.

The conditions specified for IDLE_PHY are related to running operating system and external environment test parameters. Therefore we ran this test on Ubuntu Linux and L4 Linux for every physical system in order to see the effect of the installed operating system on power consumption of physical system when running in idle state. External conditions, like test environment temperature, have also influence on physical system's power consumption therefore we tried to execute similar test in the same conditions.

Idle State Virtual Machines Power Consumption

The second test case we called IDLE_VM is specified to estimate power consumption of the physical system when running one or more VMs in idle state under different configuration parameters. With this test case we consider the two operating systems Ubuntu and L4 Linux in order to see the system's power consumption increase when certain virtualization solution is started. During first two tests no workload application was executed. The conditions specified for IDLE_VM are: number of VMs started and their parameter settings, the number of CPU cores and the size of RAM allowed for one VM.

CPU Workload Virtual Machines Power Consumption

The third test case we called CPU_VM is specified to estimate power consumption of the physical system when running one or more VMs and each VM executes certain workload. This test case tries to estimate how physical system's power consumption varies with different types of CPU workloads or bench-marks when running on one or more VMs compared to running directly on the physical system. For the workload phase of the test sequence we used different CPU and memory benchmarks: integer, memory and floating point. Every test execution was parameterized with the following settings: the number of VM instances, the running VM settings (CPU cores and memory size), and the number of simultaneous workload instances (processes or threads). Every test case was executed once on the physical system and then on the selected VMs. For every workload benchmark we logged also its performance data in order to estimate the energy efficiency for every test condition.

IO Workload Virtual Machines Power Consumption

The forth test case we called IO_VM is specified to estimate power consumption of the physical system when running one or more VMs each executing an IO workload. With this test case we try to show how virtualization influences the IO operations.

One important aspect we want to cover with this test case is hard disk I/O workload using existing disk benchmarks (therefore we may further refine this test case and name it DISKIO_VM). Other IO_VM test cases could also be implemented like USB, video, sound, etc. In our test we ran only DISKIO_VM test cases using a disk benchmark, parameterized with the following settings: the number of VM instances, the running VM settings (CPU cores and memory size), and the number of simultaneous workload instances (processes or threads).

User Applications Virtualization Power Consumption

The last test case we called USER_VM is specified to estimate power consumption of the physical system when one or more VMs are started each running the same user application. The test results are then compared with the measurements obtained when the selected user application is executed on the PM. The workload applications proposed for this test case are video player and file compressor.

4   Experimental L4 Linux Tests Results

In this section we show the results for the proposed test cases executions obtained for L4 Fiasco microkernel implementation.

4.1   Idle State Power Consumption

In this test the idle power consumption of PM was measured both for Ubuntu Linux operating system and L4 Linux microkernel. During the test execution we measured power consumption of the physical system when running in the following three conditions: (1) Ubuntu OS with X Windows graphical interface running, (2) Ubuntu OS with X Windows graphical interface stopped, and (3) L4 microkernel with X Windows graphical interface stopped.

The measured power values are shown in Table 1. It can be observed that when the X Windows system is running the physical system consumes ~0.8 W more power than when it is stopped. The measured power of the system when L4 is running is lower than the one measured when only the Ubuntu OS without X Windows interface is running. L4 has ~2% reduction in power consumption when running in idle mode.

Table 1. Idle power consumption measurements

System (1)	Ubuntu X Win	Ubuntu Console		L4 Linux Console	
AVERAGE	76.83 W	76.01 W	-1.07 %	74.50 W	-1.98 %
MINIMUM	76.60 W	75.90 W	-0.91 %	74.40 W	-1.98 %
System (2)					
AVERAGE	28.2 W	28.01 W	-0.67 %	27.50 W	-1.82 %
MINIMUM	28.1 W	27.90 W	-0.71 %	27.40 W	-1.79 %

The next presented tests results are taken for the console version of both Ubuntu OS and L4 Linux installations. We run our tests without X Windows system in order to reduce the influence of other applications over power consumption measurements.

4.2   CPU and Memory Workload Power Consumption

Within this test case we ran the same ramspeed workload with different parameters on both Ubuntu and L4 Linux systems. The first test was executed to see the influence of L4 microkernel on power consumption and memory transfer rates, therefore we ran the same workload in the same conditions on both Ubuntu and L4 systems. The power consumption measurements are shown in Fig. 2. It can be observed that power consumption of L4 microkernel during the workload phase is lower than the same phase of the test running on Ubuntu. Instead the reported performance of the benchmark for L4 is lower than the one reported for Ubuntu due to the microkernel implementation. In order to estimate energy efficiency for this workload when running on different systems we correlated the energy spent to finish the workload and the performance of the workload on host system. The obtained results are presented in Table 2.

Table 2. CPU and memory energy efficiency results

System (1)	Ubuntu	L4 Linux	
Execution time [s]	234	307	+31.19 %
Transfer rate [MB/s]	2936.47	2238.18	-23.78 %
Energy [J]	27300.60	34624.00	+26.82 %
System (2)			
Execution time [s]	246	314	+27.64 %
Transfer rate [MB/s]	2705.69	2158.31	-20.22 %
Energy [J]	11143.10	13849.60	+24.29 %

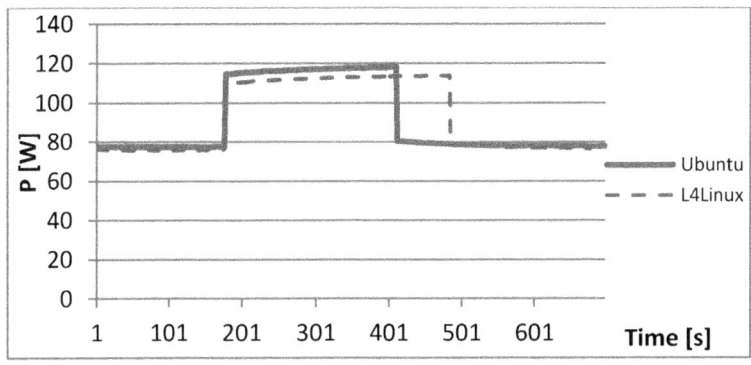

Fig. 2. CPU and memory power consumption profile

Power consumption of the system increases when running in the workload phase because the temperatures of the CPU cores increase with the execution time. In order to see how power consumption increases with workload execution time we run the same test with different number of counting times: ramspeed x 2 means double the size of workload and ramspeed x 4 specify that the workload is four times the normal ramspeed size (Fig. 3). When increasing the size of workload size the power consumption increases with almost 4 W from 112.5 W to 116.4 W on test system (2).

The current test was further extended in order to study the effect of parallelization of workload on available CPU cores. We ran two instances of ramspeed (2 x ramspeed) and four instances of the same workload (4 x ramspeed) (Fig. 3). It can be observed improvement in both performance and energy efficiency when executing multiple workload instances instead of one single instance. The obtained measurements are shown in Table 3.

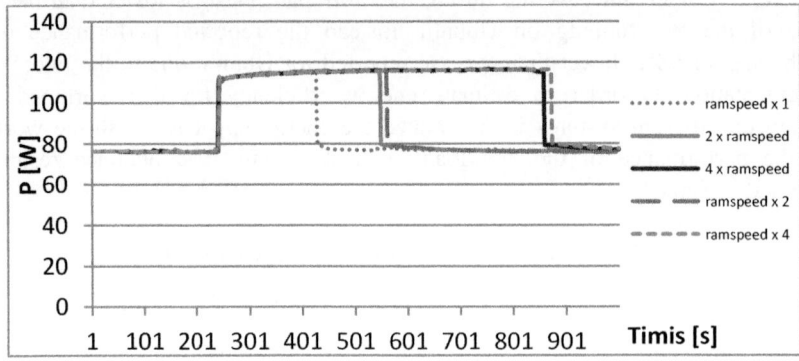

Fig. 3. Multi-tasking workload power consumption profile

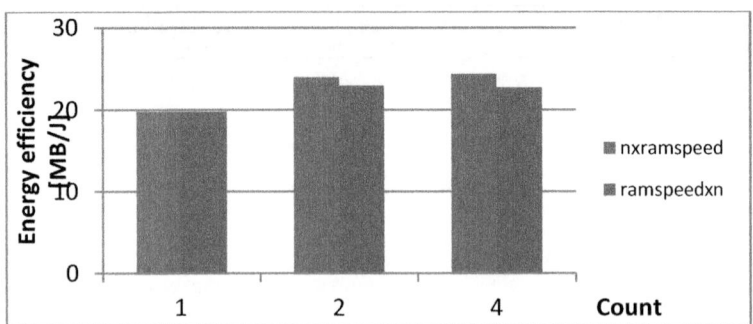

Fig. 4. Multi-tasking workload energy efficiency

Table 3. CPU and memory energy efficiency results

System (1)	2 x ramspeed	ramspeed x 2	4 x ramspeed	ramspeed x 4
Execution time [s]	306	319	618	631
Transfer rate [MB/s]	2742.03	2631.81	2807.65	2621.20
Energy [J]	35056.30	36623.60	71327.10	72818.80

In Fig. 4 energy efficiency of L4Linux ramspeed execution is shown when increasing the size of the workload. The workload is increased sequentially (called generic ramspeed x n) and parallel using multiple workload tasks (n x ramspeed). It can be observed that energy efficiency of the ramspeed workload increase when the workload size increase and when using parallelization compared with sequential execution.

4.3   IO HDD Workload Power Consumption

Power consumption of IO operations were performed using iozone hard disk benchmark. IOzone is a file system benchmark tool running on different platforms that generates and measures a variety of file operations: read, write, re-read, re-write, read backwards, read strided, fread, fwrite, random read. We executed the IO workload with different parameters on both operating systems under tests. Power measurements for one test execution are shown in Fig. 5. In our tests results we could not highlight significant differences between both performance and power consumption of IO operations with hard disk.

Table 4. Disk I/O energy efficiency results

System (1)	Ubuntu	L4 Linux	
Execution time [s]	130	131	+0.76 %
Transfer rate [ops/s]	15012	14795	-1.45 %
Energy [J]	10845.5	10777.6	-0.63 %
System (2)			
Execution time [s]	143	145	+1.40 %
Transfer rate [ops/s]	13636	13379	-1.88 %
Energy [J]	4320.72	4145.23	-0.41 %

The tests were executed using existing files of 1 GB and we used read and write operations using 1KB blocks of data. Power profiles of disk I/O test execution are presented in Fig. 5 and the same profile is obtained for both Ubuntu and L4Linux solutions. The energy efficiency is also similar on both platforms.

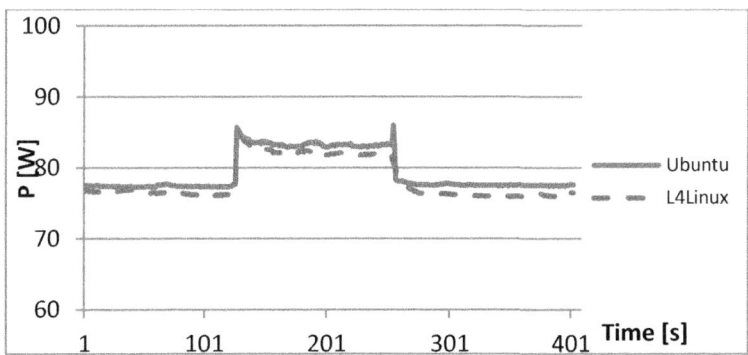

Fig. 5. Disk I/O workload power consumption profile

Table 5. Disk I/O energy efficiency results

System (1)	Ubuntu XWin	Ubuntu Console		L4 Linux Console	
AVERAGE	96.87 W	81.02 W	16.36 %	80.50 W	0.63 %
MINIMUM	95.50 W	78.80 W	17.49 %	79.10 W	-0.38 %
System (2)					
AVERAGE	36.23 W	32.02 W	11.62 %	31.12 W	-2.81 %
MINIMUM	34.50 W	31.20 W	9.56 %	30.50 W	-2.24 %

4.4   User Application Power Consumption

In order to see the power signature of real user applications running in L4 Linux we selected two applications: gzip and mplayer. We executed both applications on Ubuntu OS and L4 Linux. gzip application was used to compress and decompress a large file and its execution power consumption measurements are presented in Fig. 6. There are not important differences between file compression running on Ubuntu and L4 Linux.

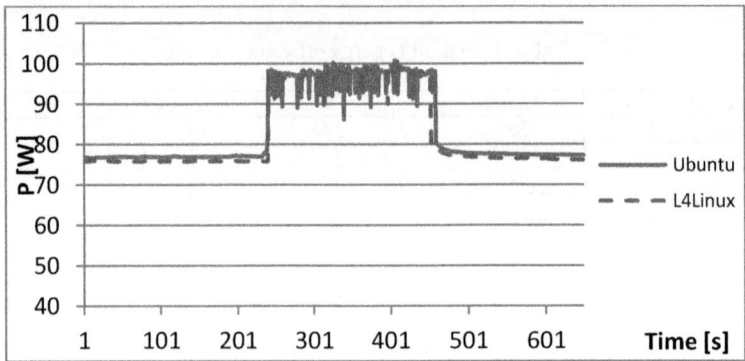

Fig. 6. Gzip compression power consumption

The last test was to run an AVI file with mplayer on three conditions: Ubuntu with X Windows system started, Ubunt without X Windows and L4 Linux without X Windows. The test results shown in Fig. 7 presents the power consumption of decoding process executed on Ubuntu and L4 Linux (Table 5).

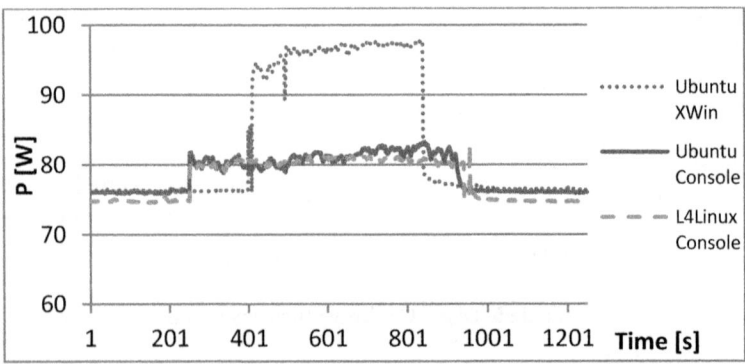

Fig. 7. Mplayer power consumption

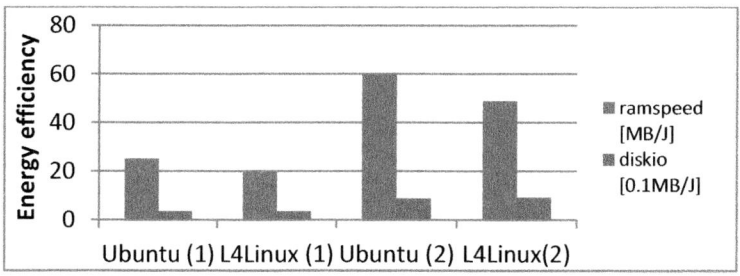

Fig. 8. Memory and I/O operations energy efficiency

5 Conclusions

This paper explores how virtualization influences the power consumption of both physical systems and virtual systems and which is the most efficient way to implement such applications. We proposed a number of test cases that can be used to evaluate power consumption and energy efficiency of virtualization systems. We run the tests on different common desktop and laptop multi-core systems.

In Fig. 9 the concluding results of energy efficiency of CPU, memory and disk I/O tests are shown. Due to the implementation particularities of L4Linux memory operations are less preformat than non-virtualization implementation therefore the energy efficiency is lower with ~25%. The disk I/O operations however are similar on both platforms Ubuntu and L4Linux. Another observation is that L4Linux implementation consumes less power than Ubuntu, when idle, on the same machine.

Acknowledgments. This work was supported by CNCSIS-UEFISCSU, project number PNII-IDEI 1009/2008. This work has been carried out in the context of the eMuCo project (www.emuco.eu), European project supported by the EU under the Seventh Framework Program (FP7) for research and technological development.

References

1. Carpenter, R.E.: Comparing Multi-Core Processors for Server Virtualization. White Paper, Intel Information Technology (2007),
   http://www.multicoreinfo.com/research/papers/whitepapers/
   multicore_virtualization.pdf
2. Dhiman, G., Marchetti, G., Rosing, T.: vGreen: A System for Energy Efficient Computing in Virtualized Environments. In: International Symposium on Low Power Electronics and Design, ISLPED 2009, USA (2009)
3. Clark, C., Fraser, K., Hand, S., Hansen, J.G., Jul, E., Limpach, C., Pratt, I., Warfield, A.: Live migration of virtual machines. In: Proceedings of the 2nd Conference on Symposium on Networked Systems Design & Implementation, NSDI 2005 (2005)
4. Tian, K., Yu, K., Nakajima, J., Wang, W.: How virtualization makes power management different. In: Proceedings of the Linux Symposium, OLS 2007, Canada (2007)

5. Chetan, S., Gautam Kumar, K., Dinesh, M.K., Abhimanyu, M.A.: Cloud Computing for Mobile World. Research Work (2010),
   http://chetan.ueuo.com/projects/CCMW.pdf
6. Nathuji, R., Schwan, K.: VirtualPower: Coordinated Power Management in Virtualized Enterprise Systems. ACM SIGOPS Operating Systems Review 41(6), SOSP 2007 (2007)
7. Verma, A., Ahuja, P., Neogi, P.: pMapper: Power and Migration Cost Aware Application Placement in Virtualized Systems. In: Proceedings of the 9th ACM/IFIP/USENIX International Conference on Middleware, Moddleware 2008, Leuven, Belgium (2008)
8. Ye, L., Lu, G., Kumar, S., Gniady, C., Hartman, J.H.: Energy-Efficient Storage in Virtual Machine Environments. In: Proceedings of the 6th ACM SIGPLAN International Conference on Virtual Execution Environments, VEE 2010, Pittsburgh, USA (2010)
9. Härtig, H., Roitzsch, M., Lackorzynski, A., Döbel, B., Böttcher, A.: L4-Virtualization and Beyond. Korean Information Science Society Review (December 2008)

Digital Holography for Security Applications

Roberto Maurizio Pellegrini[1,2], Samuela Persia[2], and Silvello Betti[1]

[1] University of Rome Tor Vergata, v.le Politecnico 1, 00133 Rome, Italy
[2] Fondazione Ugo Bordoni, v.le del Policlinico 147, 00161 Rome, Italy
{m.pellegrini,spersia}@fub.it, betti@ing.uniroma2.it

Abstract. A survey of Digital Holography (DH) and its employment in different application fields is provided. This paper reviews the main principles of the DH focusing on the optical techniques for security purposes. Recording and processing of three dimensional data, secured storage data, the use of Multimedia Sensor Network (MSN) for encrypted data transmission, and thus remote reconstruction of 3D images, are relevant examples in which DH represents an attractive solution. In this work, the state of art and major research challenges for this type of applications are shown and at the end fundamental open issue are discussed in order to outline the future research trends in this topic.

Keywords: Digital Holography, Interference, Diffraction, optoelectronics, Wireless Multimedia Sensor Network.

1 Introduction

Carefully attention has been shown in information security during the last decade. Optical information-processing techniques have proved to be a real alternative to purely electronic-processing in security, encryption, and pattern-recognition applications. This is partially due to recent advances in optoelectronic devices and components, such as detectors, modulators, optical memories, and displays. Now, in general, it is easy to transfer information from electronic to optical domains and vice versa at high speeds. In this way, it is possible to combine the advantages of both approaches to develop more efficient security applications. This fact is especially relevant when securing information codified in the form of two-dimensional (2D) or three-dimensional (3-D) images because, in these cases, optical systems are unavoidable. More deeply, with the coming of high-quality megapixel digital cameras, the creation of digital holograms of real-world objects has become viable today. They permit to reconstruct 3D images without optical equipment (e.g. eyeglasses to combine separate images, or special screen), such as stereoscopy technique, in which 3D of an objects is obtained by the *"illusion of depth"* due to the two offset images of the left and right eye of the viewer.

On the other hand a hologram is an optical element or an image that can record all information, both amplitude and phase information, present in a wave front needed to reconstruct the real 3 dimension image of an object [1].

Specifically a holographically storage image or *hologram* represents the recorded interference pattern between a wave field scattered from the object,

R. Prasad et al. (Eds.): MOBISEC 2011, LNICST 94, pp. 101–112, 2012.
© Institute for Computer Sciences, Social Informatics and Telecommunications Engineering 2012

named *wave object*, and a coherent background named *reference wave*. It is usually recorded on a flat surface, but contains the information about the entire three dimensional wave field. The process of making hologram is known as *"Holography"* and a sub area of holography is the *"Digital Holography* (DH)", that enables, thanks to the development of *Charged Coupled Devices* (CCDs), full digital recording and processing of holograms, without any photographic recording as intermediate step.

Hence the growing interest in the digital holography permits to consider new optical technique for security applications never proposed in the past, such as the secure storage of 3D data, or the pioneer remote reconstruction of 3-D image. Actually to obtain a reliable system further investigations are needed. For this scope, in this work we provide an overview of the state of art, which is the starting point for our research field. In the next future we will set up a laboratory[1] ito design a new system able to encrypt and secure transmit 3-D sensible data by using Digital Holography.

This paper is organized as follows: in Section 2, we provide a brief overview of the traditional holography, in Section 3 the main principles of the digital holography are carried out, in Section 4 we show the main holographic scenarios, while in Section 5 we pay attention to a particular type of scenarios: security scenarios takeing into account the main advantages reachable by using this technique not reachable with other technologies. Finally, in Section 6 we draw the main conclusions.

2   Holography

A hologram is the photographic record of the interference figure between the radiation scattered from an object and a coherent reference wave. Light with sufficient coherence length illuminates an object. It is scattered at the object surface and reflected to the recording medium. A second light beam, known as the reference beam, also illuminates the recording medium, so that interference occurs between the two beams. The resulting light field generates a seemingly random pattern of varying intensity which is recorded in the hologram.

2.1   Hologram Theory

Interference and diffraction effects are the theoretic concepts of the holography. Recording of the image is possible thanks to the interference effect, while reconstructing of the data is possible thanks to the diffraction effect. The holographic process is described mathematically using the description of the light propagation by the wave equation, following Maxwell formalism, in which electrical field is a vector quantity, which could vibrate in any direction, perpendicular to light propagation.

---

[1] Care of the Istituto Superiore delle Comunicazioni e delle Tecnologie dellInformazione (ISCTI) of the italian Minister Ministero dello Sviluppo Economico.

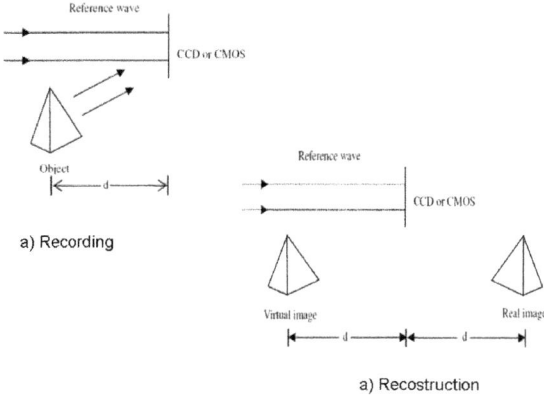

Fig. 1. Off-axis DH technique: a) Hologram recording, b) Hologram reconstruction

The superposition of two or more waves in the space is named *interference*. If we consider two monochromatic waves with equal frequencies and wavelengths the resulting intensity is the sum of the individual intensities I_1, and I_2 plus the interference term, which is:

$$2\sqrt{I_1 I_2} \cdot cos(\Delta\varphi) \tag{1}$$

this term depends on the phase difference between the waves.

The intensity distribution observed on the recording medium is the *interference figure*, and it is composed by a succession of light and dark band named *interference fringes*.

After, to reconstruct the image, diffraction effect needs to take into account. This phenomenon can be described by the Huygens' principle and mathematically analyzed by the Fresnel-Kirchhoff integral [3].

Finally, it is important to mention some drawback in the holography, for recording (*Speckles*), and reconstructing (*image distortion*) phase respectively:

- *Speckles*- This effect appears when a rough surface is illuminated with coherent light. In this situation observer sees a grainy image. The intensity of the light scattered by the surface fluctuates randomly in space, dark and bright spots appear. These spots are named *speckles*.
- *image distortion*- In the hologram reconstruction the obtained field has composed of three terms: the first term is the un-diffracted reference wave passing the hologram (*zero diffraction order*); the second term is the reconstructed object wave, forming the virtual image; and the third term generates a distorted image of the object. For off-axis holography the virtual image, the real image, and the un-diffracted wave are spatially separated.

The speckles problem can be solved by using rough surface with height variations less than the wavelength of the light, while an undistorted real image can be produced by using the conjugate reference beam for reconstruction.

3   Digital Theory

Digital holography is the technology of acquiring and processing holographic measurement data, typically via a CCD camera or a similar device. In particular, this includes the numerical reconstruction of object data from the recorded measurement data, in distinction to an optical reconstruction which reproduces an aspect of the object. Digital holography typically delivers three-dimensional surface or optical thickness data. There are different techniques available in practice, depending on the intended purpose. Among of all, in the following we report a brief description of the more extensive technique adopted:

- *Off-axis configuration* - This process was proposed for the first time in [2] where a small angle between the reference and the object beams is used. In this configuration, a single recorded digital hologram is sufficient to reconstruct the information defining the shape of the surface, allowing real-time imaging.
- *Phase-shifting holograms* - This technique uses a phase shifting algorithm to calculate the initial phase and thus the complex amplitude in any plane, e.g. the image plane. Hence with the initial complex amplitude distribution in one plane, the wave field in any other plane can be calculated by using the Fresnel-Kirchoff formulation of diffraction [4].

3.1   General Principles

In this section an overview of the theoretic principles of the DH is drawn. First of all simple evaluation of the spatial frequency requirements to obtain suitable 3-D reconstruction is provided. After two main DH techniques are taken into account in order to highlight the basic analytical concepts of both schemes.

Spatial Frequency Requirements. A CCD used to record must resolve the interference pattern resulting from superposition of the reference wave with the waves scattered from the different object points. The maximum spatial frequency to be resolved is determined by the maximum angle θ_{max} between these waves, and the wavelength λ, according to:

$$f_{max} = \frac{2}{\lambda} \cdot \frac{\theta_{max}}{2} \qquad (2)$$

The distance between the neighboring pixels of a high resolution CCD is only in the order of $\Delta x \approx 5\mu m$, thus the corresponding maximum resolvable spatial frequency can be calculated by:

$$f_{max} = \frac{1}{2\Delta x} \qquad (3)$$

Hence resolutions will be in the range of 100 line pairs per millimeter (Lp/mm). Combining the above equations we obtained that:

$$\theta_{max} \approx \frac{\lambda}{2\Delta x} \qquad (4)$$

where the approximation is valid for small angles. The distance between neighboring pixels is therefore the quantity, which limits the maximum angle between reference and object wave.

By formula it is possible to calculate the minimum distance d_{min} between object and CCD [2]. It is shown that the minimum distance linearly increases with both the dimension of the object and the CCD pixel number, while decreases with the wavelength. It means that the maximum spatial frequency has to be adapted very carefully to the resolution of the CCD.

Off-Axis Holography. The concept of digital hologram recording is illustrated in Fig.1(part a). A plane reference wave and the wave reflected from a three dimensional object placed at a distance d from a CCD or CMOS are interfering at the same surface. The resulting hologram is electronically recorded and stored. In optical reconstruction as shown in Fig.1(part b), a virtual image appears at the position of the original object and the real image is also formed at a distance d, but in the opposite side from the CCD or CMOS.

The diffraction of a light wave at CCD or CMOS is described by the Fresnel-Kirchoff integral:

$$\Gamma(\xi', \eta') = \frac{1}{\lambda} \int \int h(x,y) R(x,y) \frac{exp(-i\frac{2\pi}{\lambda})\rho'}{\rho'} dx dy \qquad (5)$$

with

$$\rho' = \sqrt{(x - \xi')^2 + (y - \eta')^2 + d^2} \qquad (6)$$

where $R(x,y)$ is the plane reference wave, $h(x,y)$ is the hologram function and ρ' is the distance between a point in the hologram plane and a point in the reconstruction plane. An undistorted real image can be produced by using the conjugate reference beam for reconstruction, i.e. to consider the conjugate plane reference wave R^* instead of R in the above equation.

In addition, it is possible to consider digitized version of the equation 5 for numerical reconstruction. Generally the numerical version of the equation can be obtain by two different approaches: reconstruction by the discrete Fresnel transformation or the Fourier Transform (FT) method; reconstruction by the convolution method (CV). For the interest reader the analytical description of both methods are in [3].

Phase-Shifting Holography. The principal arrangement for phase shifting DH is shown in Fig.2. The object wave and the reference wave are interfering at the surface of a CCD. The reference wave is guided via a mirror mounted on a *piezo-electrict transducer* (PZT). Whit this PZT the phase of the reference wave can be shifted stepwise. Several (at least three) interferograms with mutual phase shifts are recorded. Afterwards the object phase ϕ_0 is calculated from these phase shifted interferograms. The real amplitude $a_0(x, y)$ of the object wave can be measured from the intensity by blocking the reference wave.

As a result, the complex amplitude:

$$E_0(x, y) = a_0(x, y) \exp(+i\phi_0(x, y)) \tag{7}$$

of the object wave is determined in the recording (x, y) plane. By using the Fresnel-Kirchoff integral is possible to calculate the complex amplitude in any other plane [5].

The advantage of phase shifting DH is a reconstructed image of the object without the zero order term and the conjugate image. The price for this achievement is the higher technical effort: phase shifted interferograms have to be generated, restricting the method to slowly varying phenomena with constant phase during the record cycle.

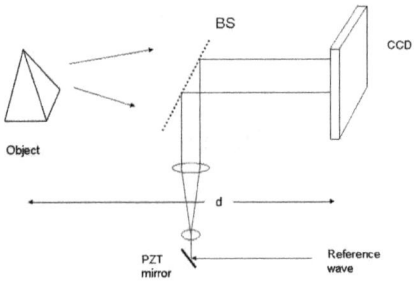

Fig. 2. Phase shifting DH, set-up

3.2 Open Issues

Holography is a 3D imaging process that instantaneously captures the volumetric information of a test object. For this purpose DH looks very promising in making holography an easy-to-use technique thanks to recent developments in megapixel CCD or CMOS cameras and the fast, efficient reconstruction algorithm by which it is now possible to record digital holograms in real time. Furthermore the use of fast computers in numerical reconstruction makes DH more flexible in terms of hologram processing.

Hence the ability of numerical evaluation of both amplitude and phase information is its main advantage over other optical imaging methods. Anyway, the limited pixel size of the commercially available digital detectors restricts the angle between the object and reference beams to a few degrees for hologram recording. Various digital methods have been proposed to solve this issue. For instance, an alternative could be perform instantaneous 3-D measurement of particle distribution as in [7], or including stereoscopic particle tracking as proposed in [8]. Actually these techniques have limitations either with volume size or particle density or the need for multiple exposures. However, in recent years Digital Holography has incurred enhancement. In addition numerical reconstruction process of particles holograms in different planes based on the Fourier transform (FT) and convolution (CV) schemes have been proposed in literature [9], to provide schemes enable the use of complex amplitude information that is inaccessible in optical reconstruction.

Future efforts in this field will be the capacity of to extraction and tracking of particles from the reconstructed images. This will permit a more deep applicability if the DH in encrypted data storage for security scenarios.

4   Digital Holography Applications

Holography can be put to a variety of uses other than recording images. In this work we are particularly interested into security applications. Anyway the application of DH can be seen in different areas:

- *Holographic Interferometry*- Holographic interferometry (HI) is a technique which enables static and dynamic displacements of objects with optically rough surfaces to be measured to optical interferometric precision (i.e. to fractions of a wavelength of light).
- *Security applications* - Holograms are used for security purposes, especially in the anti-counterfeiting field, and for optical data processing for encryption technique, hence we can distinguish:
  - *Secure Data Storage* - Holographic data storage is a technique that can store high density of information, up to become the next generation of popular storage media;
  - *Encrypted Data*   - The Digital Holography permits to encrypt and decrypt 3D data, enabling a more deep secured data transmission;
  - *Secure ID Tags* - Security holograms are very difficult to forge because they are replicated from a master hologram which requires expensive, specialized and technologically advanced equipment.
  - *CCD Image Sensors*- A revolutionary application could be the employment of DH for sensors for different purposes, especially for security: sensors will be able to perform measurements and react accordingly. In this scenario the hologram is made with a modified material that interacts with certain molecules generating a change in the fringe periodicity or refractive index, therefore, the color of the holographic reflection.
- *Dynamic Holography* - In static holography, recording, developing and reconstructing occur sequentially and a permanent hologram is produced. There also exist holographic materials which do not need the developing process and can record a hologram in a very short time. This allows one to use holography to perform some simple operations in an all-optical way (e.g. optical cache memories, image processing, optical computing).

Among of all, in this work we investigated how the DH is a suitable solution for security, and how it is possible to obtain challenger solutions not yet investigated.

5   Optical Technique for Information Security

As already mentioned DH can be applied in security field. Specifically in this section we described how it can be employed in the security field, and which advantages can be reached.

5.1   Optical Encryption and Decryption of Three Dimensional Objects

Digital Holography permits to encrypt data in 3D. The existing algorithms for encryption are based on 1D, or 2D data. Usually the 2D data are often considered as mutually orthogonal, and thus a simplification of the data processing is always justified. On the other hand, by 3D data, this approach is no more valid, and then developing of more complex algorithms are required. Consequently coding these new powerful algorithms is not trivial.

During the last decade, researchers investigated in the encryption definition field and different solutions have been provided. One of the approaches to securing information by optical means consists in the use of random phase-encoding techniques. In these methods, images or holograms are transformed, by using random phase masks, into noise distributions. Therefore, the original information remains encrypted and be recovered only by means of a random phase mask acting as the key. However, in general, the resulting encrypted data contain both phase and amplitude and, thus, must be recorded and stored holographically.

Digital holography as already mentioned is a useful technique for recording the fully complex field of a wave front. Encrypted data are stored in digital format, so that is possible to transmit, and decrypt the encrypted data digitally. For instance, in Fig.3 a secure image/video-storage/transmission system that uses a combination of double-random phase encryption and a digital holographic technique is shown.

The data are encrypted optically by the double-random phase encryption technique and recorded as a digital hologram. The optical key, that is, the Fourier phase mask, can also be recorded as a digital hologram. The encrypted data can be decrypted digitally with the hologram of the optical key.

An alternative technique for optical encryption of three-dimensional (3D) information could be the use of the Computer Generated Holograms (CGH) principle [12]. Specifically the principle of off-axis digital holography is used with single and multiple phases encoding to encrypt the 3D object. Authors showed that Multiple-Phase encoding is very secure than Single-Phase encoding but it will add a noise in case of gray-scale image compared with Black-White image, and thus more investigation are needed in that direction, in order to obtain a good level of encryption without complexity addition.

In conclusion, the 3D data encryption is possible using DH, but it requires complex processing, more than other existing technologies. On the other hand, the reliability level reachable is very high with respect to other solutions, thanks to the nature of the encrypted data: 3D data cannot be considered as the

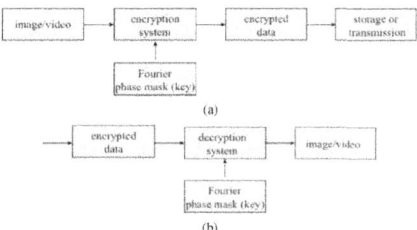

Fig. 3. Secure image/video-storage/transmission system that uses a combination of double-random phase encryption and a digital holographic technique: (a) an encryption/transmission system and (b) a receiving/decryption system.

composition of 1D data mutually orthogonal information. Consequently the developed algorithms to describe the information become complex, and not trivial to implement and simulate.

5.2 Secure Data Storage Based on Digital Holography

Digital Holography is a promising candidates for next generation of secured data elaboration. Holographic storage enables storage densities that can surpass those of traditional recording because it goes beyond the two-dimensional approaches of conventional storage technologies to write data in three dimensions. The underlying principles of holographic storage and the development of high-performance holographic recording materials are described in [10]. In addition possible solutions to obtain an encrypted optical memory system by using two 3-D keys are still open issues. Solutions are proposed in literature [11], even if more investigation are needed to obtain efficient scheme.

5.3 Wireless Multimedia Sensor Networks

Wireless Sensor Networks (WSN) have received a deep interest in research and industrial community due to many applications where they can be employed. Originally WSNs have been intended to measure physical phenomena and thus for low-bandwidth, and delay-tolerant data streams [6]. Recently, the focus is shifting toward research aimed at revisiting the sensor network paradigm even to enable delivery of multimedia content. This attitude leads to the consideration of a new sensor network concept, in which the integration of low-power wireless networking technologies with inexpensive hardware such as CMOS or CCD cameras and microphones will permit the development of a new concept of sensor networks: the *Wireless Multimedia Sensor Networks* (WMSNs). These networks are based on wireless, interconnected smart devices that enable retrieving video and audio streams, still images, and scalar sensor data [13]. A typical multimedia sensor network is shown in Fig.4. New applications are possible by using this type of networks. Among of all we remind: *Multimedia Surveillance Sensor Networks.* where sensors can be used to enhance and complement existing surveillance systems to prevent crime and terrorist attacks; or *Environmental and Structural Monitoring*, where an arrays of video sensors already are used by oceanographers can be used to determine the evolution of sandbars using image processing techniques.

To design this new sensor networks concept new paradigms are needed, among of all:

1. A real-time streaming application is more demanding than data sensing applications in wireless sensor networks primarily due to its extensive requirements for video/audio encoding. The limitations of the sensor nodes require video coding/compression that has low complexity, produces a low output bandwidth, tolerates loss, and consumes as little power as possible.
2. The area where multimedia streaming applications are different from other applications in wireless sensor networks is in the usability of encryption techniques to ensure confidentiality. We remind that in a wireless sensor network, the public key cryptography schemes are not suitable because of their high power and computation requirements. On the other hand, standard symmetric encryption schemes, such as DES and AES, are commonly used. However, these schemes are unsuitable for multimedia data. Multimedia data is generally larger in size and use of these symmetric encryption schemes has memory and computation requirements that are unsupportable by the sensor nodes.
3. To supporting multimedia traffic, a new concept of the transport layer is required, because in the traditional WSN for data transmission, the notion of end-to-end packet delivery reliability is in most cases unnecessary.

The use of holographic technique permits to solve the issues related to the multimedia transmission than the other solutions. For instance image/video encoding and transmission by DH, as described in the previous sections, will support less resource demanding than other technique. Hence DH can be used in this type of networks by considering the transmission of secure digital hologram trough the network. The CCD sensors will be connected into the distributed network for the transmission encrypted data. In contrast to Radio Frequency (RF), optical devices are smaller and consume less power; reflection, diffraction, and scattering from aerosols help distribute signal over large areas; and optical wireless provides freedom from interference and eavesdropping within an opaque enclosure. Optics can accommodate high-bandwidth transmission of multimedia without meet harmful effect due to interference of radio propagation. These motivate use of optical wireless as a mode of communication in sensor networks, and the capability of delivering high level of data due to the optical data storage not reachable with other technologies. In conclusion future researches will require to define solutions able to joint power optimization and efficient channel coding to design a new wireless sensor network for multimedia application. Actually due to the lack of comprehensive comparison studies and test-bed implementations, at the moment it is no possible to provide quantitative comparisons.

In addition, with DH it will be possible to reconstruct image remotely, without optics equipment (e.g. lens, screen) because each pixel, compounding the image, have got both amplitude and phase information needed to reconstruct the original object image. In this context, Optical Holography permits to reconstruct image in the laboratory, while by using DH it is possible to reconstruct the image remotely through a complex opto-electronic system composed by fibers, and electro-optics devices. In this way, collected images by Wireless Multimedia Sensor Networks, will be available and correctly reconstructed for instance $1000Km$ far away. Actually the system will require high costs due to both complex opto-electric devices, and powerful computers to carry out data processing. This is the pioneer system that we intend to investigate in the next years. For this scope we will set up a laboratory in the next future to perform measurements in this field. Specifically our aim is to define the guidelines of the system with restricted costs, as low as possible, in order to consider the system feasibility.

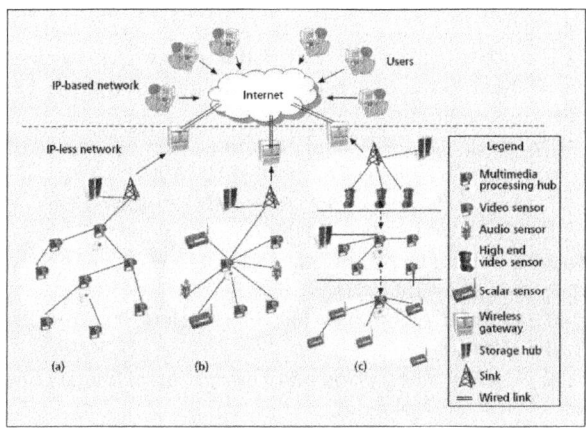

Fig. 4. Wireless Multimedia sensor Network: an architecture

6   Conclusions

In this work we discussed the state of art of research on Digital Holography, paying carefully attention in the possible application fields. Among of all, security applications have been taken into account due to the peculiarities of the Digital Holography which permit to obtain solutions not reachable with other technologies.

Indeed three main security applications have been discussed: *Secure Data Storage, Encryption Data*, and *Wireless Multimedia Sensor Networks*. By using DH is possible to storage big size of data in secure way, encrypt 3D data in more efficient manner, and finally record and transmit 3D measurements by using multimedia sensor network. The latter application is a new challenge in digital 3D transmission information, because it is based on new opt-electronic system enable to record data and to reconstruct 3D image remotely. It will permit to define a system never proposed in the past, and thus transmit sensible complex 3D data in secured manner.

References

1. Gabor, D.: A new microscopic principle. Nature 161, 777–778
2. Leith, E.N., Upatnieks, J.: Reconstructed wavefronts and communication theory. Journ. Opt. Soc Amer 54, 1295–1301
3. Schnars, U., Jueptner, W.: Digital Holography. Springer Edition (2005)
4. Skarman, B., Becker, J., Wozniak, K.: Simultaneous 3D-PIV and temperature measurements using a new CCD-based holographic interferometer. Flow Meas. Instrum. 7(1), 1–6
5. Seebacher, S.: Anwendung der digitalen Holografie bei der 3D-Form- und Verformungsmessung an Komponenten der Mikrosystemtechnik. Univeristy Bremen publishing house, Bremen (2001)
6. Akyildiz, I.F., et al.: A Survey on Sensor Networks. IEEE Communication Magazine (August 2002)
7. Fournier, C., Ducottet, C., Fournel, T.: Digital holographic particle image velocimetry: 3D velocity field extraction using correlation. Journal of Flow Visualization and Image Processing 11, 1–20
8. Virant, M., Dracos, T.: 3D PTV and its application on Lagrangian motion. Measurement Science and Technology 8, 1539–1552 (1997)
9. Ohmi1, K., Panday, S.P., Joshi, B.: Digital Holography Based Particle Tracking Velocimetry. In: FLUCOME 2009 (2009)
10. Dhar, L., Hortelano, E.: The Next Generation of Optical Data Storage: Holography, RADTECH Report (November, December 2007)
11. Matoba, et al.: Optical Techniques for Information Security. Proceedings of the IEEE 97(6) (June 2009)
12. Kishk, S., et al.: Optical Encryption and Decryption of Three Dimensional Objects by Computer Generated Holograms, ???
13. Akyildz, I.F., Melodia, T., Chowdury, K.R.: Wireless Multimedia Sensor Networks: A Survey. IEEE Wireless Communications (December 2007)

Formal Security Analysis of OpenID with GBA Protocol

Abu Shohel Ahmed[1] and Peeter Laud[2]

[1] Ericsson Research
[2] Cybernetica AS and Tartu University

Abstract. The paper presents the formal security analysis of 3GPP standardized OpenID with Generic Bootstrapping Architecture protocol which allows phone users to use OpenID services based on SIM credentials. We have used an automatic protocol analyzer to prove key security properties of the protocol. Additionally, we have analyzed robustness of the protocol under several network attacks and different threat models (e.g., compromised OP, user entity). The result shows the protocol is secure against key security properties under specific security settings and trust assumptions.

1 Introduction

In today's world, digital identity [26] of a person is important for accessing online services. At present, most online (i.e., application) service providers maintain their own identity management systems. This approach requires the user to register and authenticate separately for each online services. OpenID is one such solution which can overcome these problems. OpenID (OpenID 2.0 [22]) allows the user to control his identity information and it also allows access to several online services (i.e., web sites) with a single identifier [23].

Although OpenID provides an easy way for services to identify the user, the usability of OpenID is questionable [9]. The user is still required to type username and password (most common authentication mechanism) at OpenID provider to prove his identity. Moreover, from the security perspective, humans are not good at choosing and remembering strong passwords [18]. All these problems suggest that current OpenID implementations are not user friendly and secure for smart phone platforms.

For a long time, telecommunication companies have been successfully dealing with large subscriber identity systems. Telecommunication companies can easily assert identity attributes to a real person or against a subscriber identity module (SIM). Using this unique property, 3rd Generation Partnership Project (3GPP) has defined an internetworking combining OpenID with generic bootstrapping architecture (GBA). GBA is a mechanism for generating shared keys between a smartphone and a mobile network operator (MNO) [1]. The generated shared key is used to authenticate the user to the OpenID provider. This process poses a number of interesting security questions. We have addressed those questions

R. Prasad et al. (Eds.): MOBISEC 2011, LNICST 94, pp. 113–124, 2012.

by performing a formal security analysis of the protocol. Our analysis concludes that the protocol is secure within the specific trust relationships of participating parties.

To date, a number of security analyses of OpenID have been performed. Existing analyses concentrate mostly on the (lack of) protection of the HTTP traffic between the user, the relying party, and the OpenID provider. Sovis et al. [24] have found that if the HTTP 302 redirection performed by the relying party and the OpenID provider is onto HTTP (not HTTPS) endpoints, then these redirections can be changed later by the attacker. Urueña et al. [25] discuss the information leaks to third parties through HTTP headers. Both Sovis et al. and Feld and Pohlmann [12] discuss attacks based on lack of authentication of the OpenID provider. Same kinds of attacks are also discussed by Lindholm [17] who has performed, by our knowledge, the only tool-supported analysis of OpenID so far, using the AVISPA protocol analyser [5]. AVISPA is also used to analyze security properties of EAP AKA [27], which similar to GBA is based on UMTS authentication and key agreement protocol. In this paper, we are interested in the interference of OpenID and GBA, hence we do not consider these kinds of issues and model the protocols in a way that in the OpenID protocol, the necessary authentication of parties has been performed. In this paper, section 2 describes the inter-networking of OpenID with GBA and base security assumptions (i.e., honesty of parties and security of channels). Section 3 presents the formalization of the protocol using Proverif [8]. Section 4 presents verification results from our analyses. Finally, section 5 presents several scenarios based on modified base security model.

2   OpenID with GBA Protocol

This section describes the inter-networking of OpenID with GBA according to TR 33.924 [2]. The OpenID with GBA protocol consists of the following parties.

- The user U wants to authenticate to a Relying party (RP) using OpenID with GBA protocol. Our analysis assumes implicit users.
- The user entity (UE) (e.g., a client application combined with GBA modules in a Smartphone) is actually used to access the RP. Although, the original protocol [2] contains a separate GBA client module and a UICC[1] module, we consider them as a single user entity. Initially, both the UICC and the HSS[2] share a master secret key K.
- The RP contains resources to which the user wants to connect.
- NAF-OP is a combination of OpenID Provider and Network Application Function. The NAF-OP has a secret key sk_O which has a public counterpart pk_O that is bound to the NAF-OP by $cert_O$. The NAF-OP acts as a trusted party for the RP.

---

[1] Universal Integrated Circuit Card.

[2] Home Subscriber System.

- The Bootstrapping server (BSF) is a central party of the protocol. It acts as a trusted party and mediates the application specific user session key between the user entity and the NAF-OP. The BSF has a secret key sk_B which has a public counterpart pk_B that is bound to the NAF-OP by cert_B.
- The Home Subscriber Server (HSS) shares a master secret with the UICC. The HSS has a secret key sk_H which has a public counterpart pk_H that is bound to the HSS by cert_H. The BSF and the HSS maintain a both way secure channel.

Besides, a Certificate Authority (CA) exists to issue certificates for NAF-OP, BSF, and HSS. There are also means to verify the validity of the certificate.

2.1 Protocol Description

The protocol is depicted in Fig. 1. The protocol session is initiated by the user (not shown) deciding to access resources from the Relying party through the user entity. The user entity provides the user-supplied identifier O_u to the Relying party. The RP normalizes and discovers [22] an identity provider based on the O_u. After the discovery, the RP and the NAF-OP establish a shared key using Diffie-Hellman (DH) association. The DH is performed over a server authenticated TLS tunnel [10].

The RP sends a redirection message (includes claimed id O_i, relying party identity O_{rt}, provider identity OP_{End}, DH association key H_{DH} to the NAF-OP through the user entity. The redirection message between the UE and NAF-OP is sent through a server authenticated TLS tunnel. The NAF-OP uses HTTP digest authentication [13] for user authentication. According to HTTP authentication, the NAF-OP sends a HTTP digest challenge to the UE containing a nonce N_0 and NAF_{realm}. To reply this challenge, the UE requires a valid user name and password. This leads the UE to start the bootstrapping process.

Bootstrapping is a four step process based on HTTP digest AKA [20]. At first, the UE starts bootstrapping by sending the $IMPI$[3] to the BSF. The BSF, in turn, creates a two way secure channel with the HSS. The HSS contains the shared master secret key K and a running sequence number SQN for each $IMPI$. The HSS generates $RAND$: a random number; $XRES$: a number generated from one way hash function [6] using K, $RAND$, and SQN; CK: a generated session key for confidentiality; IK: a generated session key for integrity; $AUTN$: contains encrypted sequence SQN, authentication management field AMF, and a message authentication code (MAC) [21]. The generated $RAND$, $XRES$, CK, IK, and $AUTN$ are sent to the BSF using the secure channel. Second,the BSF sends a challenge to the UE containing $RAND$, and $AUTN$. The UE generates RES, CK, IK, and MAC based on its own shared master key K and sequence number SQN'. At this stage UE performs two equivalence tests: 1) it checks the UE sequence number is big enough from the HSS sequence number. 2) it also checks the equivalence of MAC received from the BSF with its own MAC. The UE authenticates the BSF based on the results of these two tests. Third,

---

[3] IP Multimedia Private Identity.

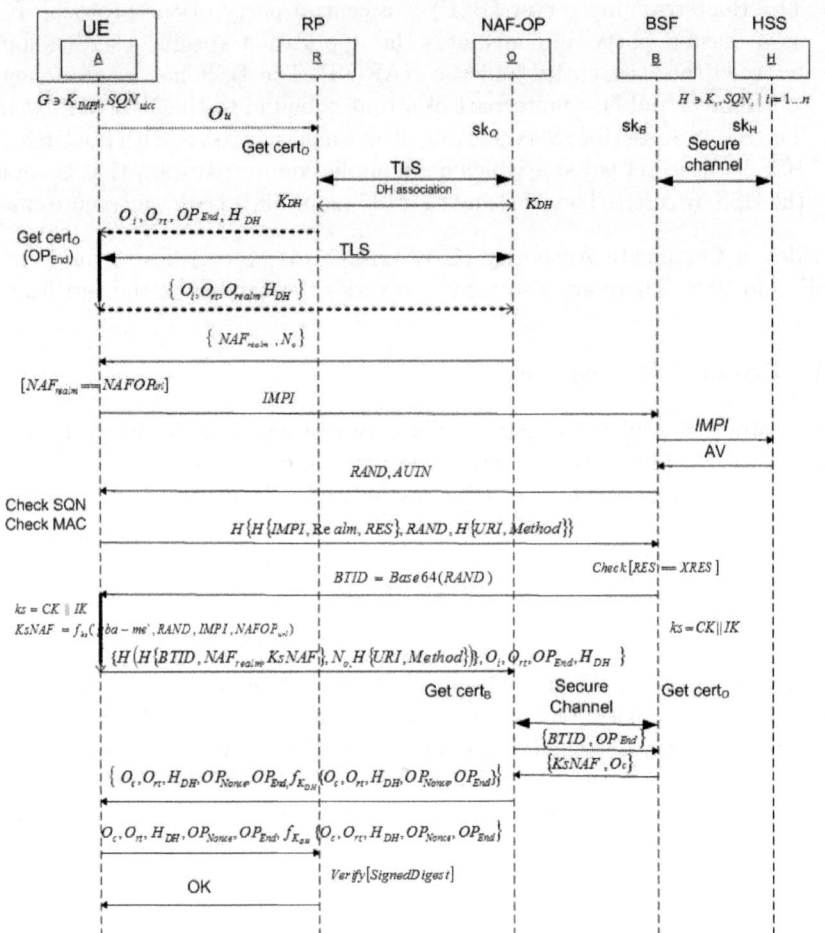

Fig. 1. OpenID with GBA protocol

the UE sends a HTTP digest response (with RES as password) to the BSF. The BSF validates the response against the previously stored $XRES$ value. The BSF authenticates the UE if only if the received RES and generated $XRES$ are equal. Upon successful UE authentication, the BSF generates a bootstrap key identifier ($BTID$) and a new temporary shared key Ks (Ks is a combination of CK and IK). Finally, the BSF sends $BTID$ to the UE. The integrity of $BTID$ is protected using HTTP digest parameter $qop = auth - int$. The UE stores the $BTID$, temporary shared secret key Ks (i.e., generated from CK and IK). At this stage, both the BSF and the UE are mutually authenticated and share a new master session key Ks.

To complete the HTTP authentication with the NAF-OP, the UE generates an application specific key $KsNAF$ using Ks and NAF-OP identity. The UE generates a HTTP digest response for the NAF-OP based on $BTID$ as a user name and $KsNAF$ as a password. The NAF-OP verifies this reponse in consultation with the BSF. Finally, the NAF-OP redirects the UE to the RP with a signed (i.e., using DH key) assertion. The UE is allowed to access RP resources based on the assertion (e.g., signed digest) from the NAF-OP.

2.2 Constraints

OpenID with GBA can be deployed in several security contexts (e.g., messages can communicate within server authenticated tunnel or in non-tunneled mode). Our analysis is based on common deployment scenarios: the UE and the RP communicates in the non-tunneled mode while the communication between the UE and the OP, and the communication between the RP and the OP are within tunneled mode. The communication between the UE and the BSF is in non-tunneled mode while OP to BSF and BSF to HSS use two way secure channels. In addition, the analysis is applicable for GBA_ME [15, p. 45] which is the most common variant of GBA. The model only focuses on the authentication steps of GBA_ME. This implies the synchronization of SQN in GBA_ME has been performed beforehand. Additionally, our model is valid for stateful OpenID where an association (i.e. D-H) is performed between the RP and the OP to verify the subsequent protocol messages.

2.3 Base Security Model

We have made the following assumptions about the security of various channels and entities as our base security model.

1. There will be several users (i.e., user entity) and RP sites, some of the RP sites are under adversarial control.
2. The UICC and the GBA modules reside within the user entity that is under the control of the respective user. All these modules are trusted in our base security model.
3. The channel between the user and his user entity is secure. We assume the client is trustworthy and only trustworthy client can access the GBA and the UICC module.
4. There are multiple honest NAF-OP servers.
5. There is only one honest BSF, and one HSS server controlled by the mobile network operator. The channel between them is secure.

We are interested in the secrecy properties of master key K shared between the UICC and the HSS, temporary session key Ks shared between the BSF and the UE, and application session key $KsNAF$ shared between the UE and the NAF-OP. We are also interested in the TLS keys agreed between honest users and servers. Additionally, we are interested in the correspondence properties e.g., if a server thinks that it has communicated with a client controlled by a user using a key k and the user is honest, then the user also thinks that it has communicated with the server in a session where the client application is using the key k *(integrity of servers)* [16]. The same principle must hold if we swap the role of the user and the server *(integrity of clients)*.

3   Formalization in ProVerif

ProVerif [7] is a mature protocol analyzer based on the Dolev-Yao [11] (or: perfect cryptography) model. This model, which has been extremely fruitful for cryptographic protocol construction and analysis, abstracts from the actual cryptographic algorithms by stating that only an enumeration of operations on messages — the actual invocations of algorithms — make sense, both for the honest parties and the adversary.

The analyzer takes as input a protocol specified in a dialect of the applied pi-calculus [3] and a specification of a security property. It verifies whether the protocol satisfies this property, answering either correctly or inconclusively (because of undecidability of protocol verification). The process language contains primitives for sending and receiving messages over channels, generating new names as keys, nonces, channel names etc., constructing and destructing messages, branching, sequential and parallel composition, and replication. The security property can either be a secrecy property (the attacker is unable to deduce a certain value), a correspondence property (certain parts of the protocol can be executed only after other parts have been executed), or a certain process equivalence. The verification proceeds by translating the protocol into a set of Horn clauses (possibly abstracting from the actual behaviour of the protocol) and then applying a specialized inference engine on it. Below we describe our model of OpenID with GBA protocol.

Secure Channels. In the OpenID with GBA protocol, certain parties have secure channels between them. To model those, we need a channel name that only these parties know. We obtain such name for two parties as $\mathsf{tochannel}(sk_1, pk_2) = \mathsf{tochannel}(sk_2, pk_1)$ where sk_i is the secret key of the i-th party, $pk_i = \mathsf{pk}(sk_i)$ is the public key corresponding to it, tochannel and pk are term constructors, and the equation is stated to hold (ProVerif can model certain equations, including this one). The public keys are bound to names by means of *certificates*.

TLS Handshake. Certain parties exchange messages over confidentiality-providing TLS tunnels where one end (the server) is authenticated. We do not want to analyze the TLS protocol in this work, because its security properties have been thoroughly studied [14,19]. To model the tunnel, we again construct a new channel name. It is done by the client C sending the server S a nonce N encrypted under the public key of S, after which the channel is defined as $\mathsf{tlschan}(N, pk_S)$, where tlschan is a public constructor. The server is authenticated by its ability to decrypt and find N. The honest parties will construct secure channels and TLS tunnels only after checking the *certificate* of the other party, thereby excluding man-in-the-middle attacks.

Certificates and Identification. The names of the parties of the protocol are bound to their public keys by means of certificates. Certificates are public messages, signed by a trusted party — the certificate authority (CA). While detailed modeling of certificates and CAs is possible in ProVerif, we want to avoid it, and thus abstract a certificate as a message $\mathsf{cert}(X, pk_X, role)$, where

X is the name of the party, pk_X its public key, and the third argument shows in which role (NAF-OP, or BSF, or HSS) this public key is intended to be used. The issuing and publication of certificates of honest parties are modeled by constructing that certificate inside the process of that party and outputting it on a public channel. The constructor cert is *private* — the attacker cannot cannot construct such messages itself. On the other hand, the destructors, giving access to the components of a certificate, are public.

To model dishonest parties inside the system, the attacker needs a means to obtain certificates to public keys where it knows the corresponding private key. Hence we include a simple "CA" in our analyzed process. This CA receives a public key from the attacker and issues certificates for it. The name appearing in the certificate is chosen by the CA, not the attacker, hence the attacker cannot masquerade as an honest party.

Symmetric master keys shared by the BSF and each of the user devices are modeled similarly. There is a private constructor impi2key that transforms IMPI into the master key. Similarly to certificates, the adversary can get (IMPI,key)-pairs from the simple CA.

The Whole System. After the abstractions described above, the modeling of the system is quite straightforward. We have processes for the user equipment (client application, GBA and UICC are all modeled in a single process), NAF-OP, RP, BSF, and HSS running in parallel, and replicated (except BSF and HSS) to model several parties of the same kind. All processes, except the user equipment and the RP, start by generating a name for themselves and creating their public and private key pair. They will publish the certificate binding their name, their role and their public key, and will then continue with the protocol sessions [4].

Note that in our model, RPs do not have certificates. The users will thus communicate with them over public channels (i.e. using HTTP, not HTTPS). The lack of RP certificates also means that they may be under adversarial control. It is noteworthy that the results we report in the next section are valid in this setting. On the other hand, the lack of RP certificates means that the users can never be sure about the identities of RPs and the security properties mentioning those cannot be verified. We do not consider this the weakness of our model — using HTTPS (or: TLS handshake) would be the standard way for the user to verify the identity of RP and the properties of TLS imply that in this case, the user can be sure who he has connected with.

The users also do not have certificates. However, the users have IMPIs and, through the private mapping impi2key, the user shares symmetric keys with the BSF.

4   Verification Results

In this section, we report the security properties we are interested in, the way we are modeling them in ProVerif, and the verification result. We perform the formal analysis in two phases: first only for bootstrapping process, second, the combined OpenID with GBA protocol (which, according to our knowledge, has not been subject of such formal analysis before).

4.1   BootStrapping

Secrecy. We are interested in the secrecy of long-term keys, as well as the short-term keys *Ks* and *KsNAF*. In the Dolev-Yao model, we model this property as the attacker's inability to obtain the term corresponding to this key. In ProVerif, such queries are straightforward to state. We obtain that all keys initially shared or constructed during a protocol session between an honest user and an honest BSF remain secret.

Authentication. We want to make sure that if the BSF thinks it has completed a session with the user U (and associated the value *BTID* with the IMPI of U), then the user U actually has participated in a protocol session with the BSF. This is a typical correspondence property and it can be easily specified in ProVerif. The protocol code may contain statements "**event** M", where M is a message. The execution of this statement has no effect on the protocol run, but the occurrence of the event M can be recorded. ProVerif can answer queries regarding the order in which the events occur.

To model this property, we add events *BSFEnd(impi)* in the end of the process modeling a session for BSF, and *UICCBegin(impi)* in the beginning of the process modeling a session for UICC. We ask ProVerif whether each *BSFEnd(i)* must be preceded by a *UICCBegin(i)* for any term i. The answer is negative because the attacker may also be known as a user of the system to the BSF, and the attacker does not emit the event *UICCBegin(...)* in the beginning of its session. So we check whether *impi* is the IMPI of an honest participant before emitting *BSFEnd(impi)*. We do this by the common technique of emitting the honest IMPIs on a private channel (this is added to the user process) and trying to receive *impi* from this channel before emitting the event. Now ProVerif states that the correspondence property holds — each *BSFEnd(impi)* for an honest *impi* is preceded by *UICCBegin(impi)* in all traces. The correspondence is even injective — different *BSFEnd*-s are preceded by different *UICCBegin*-s.

We also want to make sure that if the UICC thinks it has completed an authentication session, then the BSF also thinks the same and they also agree on the sequence numbers (and other parameters) they used. To model this property, we add the event *BSFBegin(impi,sqn)* before the last message sent by the BSF to the UICC, and add the event *UICCEnd(impi,sqn)* after the UICC has received this message. We want each *UICCEnd*-event to be preceded by a *BSFBegin*-event with the same parameters. ProVerif can verify that the correspondence property holds, but it cannot show injective correspondence. This is caused by our inability to model that a sequence number can be used only in a single session, and cannot be repeated. From the non-injective correspondence, and from the non-repeatability of sequence numbers we can deduce that in reality, there is an injective correspondence between those events.

4.2   OpenID with GBA

Secrecy. The critical resource is application specific key *KsNAF*. ProVerif verifies that application specific key is still not revealed to the attacker.

Authentication. The main property that we are interested in, is that when a protocol session is finished by RP with a user identified by UID, then there is a user that also participated in this session and this user has a legitimate claim of being identified as UID. Because of the multitude of identities and authorities, the precise property is not trivial to state.

First we note that OpenID with GBA protocol, as we have specified it here, is a *pure authentication protocol* — as the user and the RP are not establishing any common secrets, the only information that the RP gets from the protocol is that a user with a legitimate claim to the identifier UID was alive during the protocol session and used the same OpenID provider that RP thinks it used. Hence we put the event $RPEnd(RPID, UID, OPID)$ to the end of RP's process for a protocol session.

The user is using the identifier UID with RP and OP, and the identifier *impi* with the BSF. The identifier *impi* will be bound to the transient value BTID by the BSF. The OP will receive BTID from both the BSF and the user. The OP also received UID from the user and can bind it with BTID. Hence we come up with the requirement that if the RP accepts the UID then a user has started a session as UID and these bindings have been done. The property can be modeled by adding event $UserEnd(RPID, impi, UID, OPID)$ to the end of the user process (before sending the last message to the RP), adding event $OPEnd(UID, BTID, OPID)$ to the end of the NAF-OP process (before sending the last message to the user) and adding event $BSFEnd(BTID, impi)$ to the BSF process (before sending the last message to the OP). Our check with ProVerif shows that the event $RPEnd$ must be preceded by all of these three events with matching parameters; the correspondence is even injective. As before, we emit the $RPEnd$-event only if the user and the OP are honest.

Arguably, the property described in previous paragraph does not give strong guarantees to the RP. It ensures that whenever RP accepts a user as UID, there has been a user that has used UID and is correctly identified to the BSF. Two different occurrences of the same UID may in principle come from different users.

We thus also consider the scenario where the RP learns user's real identity (derived from *impi*) from the OP at the end of the protocol [22]. In this case the authentication property is straightforward to model — each event $RPEnd(RPID, UID, OPID)$ must be preceded by $UserEnd(RPID, UID, OPID)$, where UID is the real identity of the user. ProVerif states that this correspondence property does indeed hold.

Anonymity. We may be interested in the adversary not learning the real identities of the users (i.e. their *impi*-s) participating in the protocol. Such anonymity is a type of process equivalence where the process with a random *impi* is indistinguishable (to the adversary) from the process where *impi* has been selected by the adversary. Such *noninterference* queries are supported by ProVerif, although the treatment is incomplete — ProVerif checks the stronger property of the two processes evolving in lock-step.

For the case where RP does not learn the real identities of users (as the RP may also be adversarial), we have checked whether the process is noninterferent with respect to *impi*. Quite clearly, it is not — *impi* is sent in clear text from the user to the BSF. But when we move this transmission to a private channel, ProVerif succeeds in proving the noninterference.

5   Security against Malicious Participants

The inter-networking of OpenID with GBA, as modeled in section 2.1 and based on the base security model of section 2.3 is proved secure using ProVerif. However, the base security model only reflects an ideal situation. If we modify the security model certain things can go wrong.

5.1   BSF under Adversarial Control

BSF acts as an identity provider both for the user and the NAF-OP (i.e., *BTID* acts as a transient user identity for both). This means the protocol runs with an honest BSF. Quite clearly, an adversarial BSF can break the protocol. We have modeled this situation by distributing fake certificates of BSF (allowing the adversary to obtain a certificate for the role BSF). In this case, both the user and the OP lack knowledge of the honest BSF. As a result, any rogue client can convince the OP and subsequently the RP that everything is fine.

5.2   NAF-OP under Adversarial Control

The NAF-OP provides an assertion to the RP that a user controls a particular identity. Therefore, an honest NAF-OP is required to prove the validity of the user claim. We modeled a NAF-OP under adversarial control by distributing fake certificates of the NAF-OP. Under such circumstances, the malicious NAF-OP can bypass the user authentication process with the BSF. In addition, it can trick the RP to accept invalid users while rejecting the valid user from gaining access to the service. Another form attack could be, the user is confused about the identity of NAF-OP. This can lead the user to pass user credentials to a malicious NAF-OP. The malicious NAF-OP, in turn, tries to use the user credential to login to a different service. This type of phishing attack is not severe against OpenID with GBA because the user only provides site specific key to the malicious NAF-OP. This meas the attacker cannot use the same user credential for login to a different service.

5.3   Attacker Partially Controls the User Entity

The user entity consists of three modules: UICC, GBA client and an application. The severity of the attack depends on the level of control attacker has over these modules [15]. First, the UICC module can be considered secure because the

UICC card securely contains the the shared master key K. Second, the GBA client contains a temporary master session key Ks. Exposure of this key enables the attacker to login any GBA supported NAF-OP within a particular session. Third, if the attacker controls the application (e.g., browser) then the application specific key $KsNAF$ will be exposed. Exposure of $KsNAF$ enables the attacker to authenticate in a particular NAF-OP within a particular session. We see that the control of any of these three modules directly leads to a successful attack, justifying our coarse-grained model of the user entity.

6   Conclusion

The paper reports the formal specification and security properties of a non-trivial, practical protocol — OpenID with GBA. The security analysis has been performed based on specific security settings and trust relationships among participating entities. The security analysis suggests that the inter-networking of OpenID with GBA is secure in the ProVerif model. The protocol maintains both the secrecy and the correspondence properties. However, the protocol breaks under strong adversarial models. In our security model, one of the main concerns is the identity of the RP due lack of RP certificates. In addition, we have not considered the weaknesses of cryptographic algorithms.

References

1. 3GPP TS 33.220 - Technical Specification Group Services and System Aspects; Generic Authentication Architecture (GAA); Generic bootstrapping architecture (2007), http://www.3gpp.org/ftp/Specs/html-info/33220.htm
2. 3GPP TR 33.924 - Identity management and 3GPP security interworking; Identity management and Generic Authentication Architecture (GAA) interworking (2009), http://www.3gpp.org/ftp/Specs/html-info/33924.htm
3. Abadi, M., Fournet, C.: Mobile values, new names, and secure communication. In: POPL, pp. 104–115 (2001)
4. Ahmed, A.S., Laud, P.: ProVerif model files for the OpenID with GBA protocol (2011), http://research.cyber.ee (last accessed March 30, 2011)
5. Armando, A., Basin, D., Boichut, Y., Chevalier, Y., Compagna, L., Cuellar, J., Drielsma, P.H., Heám, P.C., Kouchnarenko, O., Mantovani, J., Mödersheim, S., von Oheimb, D., Rusinowitch, M., Santiago, J., Turuani, M., Viganò, L., Vigneron, L.: The AVISPA Tool for the Automated Validation of Internet Security Protocols and Applications. In: Etessami, K., Rajamani, S.K. (eds.) CAV 2005. LNCS, vol. 3576, pp. 281–285. Springer, Heidelberg (2005)
6. Bellare, M., Canetti, R., Krawczyk, H.: Keying Hash Functions for Message Authentication. In: Koblitz, N. (ed.) CRYPTO 1996. LNCS, vol. 1109, pp. 1–15. Springer, Heidelberg (1996)
7. Blanchet, B.: An efficient cryptographic protocol verifier based on prolog rules. In: Proceedings of the 14th IEEE Workshop on Computer Security Foundations, CSFW 2001, Washington, DC, USA, pp. 82–96. IEEE Computer Society (2001)
8. Blanchet, B.: An Efficient Cryptographic Protocol Verifier Based on Prolog Rules. In: 14th IEEE Computer Security Foundations Workshop (CSFW-14), Cape Breton, Nova Scotia, Canada (2001)

9. Dhamija, R., Dusseault, L.: 7 Flaws of Identity Management: Usability and Security Challenges. In: IEEE Security & Privacy, vol. 6, p. 24 (March 2008)
10. Dierks, T., Rescorla, E.: The Transport Layer Security (TLS) Protocol Version 1.2. RFC 5246 (August 2008)
11. Dolev, D., Yao, A.C.: On the security of public key protocols. Tech. rep., Stanford, CA, USA (1981)
12. Feld, S., Pohlmann, N.: Security analysis of OpenID, followed by a reference implementation of an nPA-based OpenID provider. In: Information Security Solutions Europe (ISSE) Conference, Madrid, Spain (2008)
13. Franks, J., Hallam-Baker, P., Hostetler, J., Lawrence, S., Leach, P., Luotonen, A., Stewart, L.: Http authentication: Basic and digest access authentication (1999)
14. Gajek, S., Manulis, M., Pereira, O., Sadeghi, A.-R., Schwenk, J.: Universally Composable Security Analysis of TLS. In: Baek, J., Bao, F., Chen, K., Lai, X. (eds.) ProvSec 2008. LNCS, vol. 5324, pp. 313–327. Springer, Heidelberg (2008)
15. Holtmanns, S., Niemi, V., Ginzboorg, P., Laitinen, P., Asokan, N.: Cellular Authentication for Mobile and Internet Services., 1st edn. Wiley Publishing Inc. (2008)
16. Laud, P., Roos, M.: Formal Analysis of the Estonian Mobile-ID Protocol. In: Jøsang, A., Maseng, T., Knapskog, S.J. (eds.) NordSec 2009. LNCS, vol. 5838, pp. 271–286. Springer, Heidelberg (2009)
17. Lindholm, A.: Security Evaluation of the OpenID Protocol. Master's thesis, Royal Institute of Technology (KTH), Stockholm, Sweden (2009)
18. Menezes, A.J., van Oorschot, P.C., Vanstone, S.A.: Handbook of Applied Cryptography, 5th edn., vol. 1, ch.10. CRC Press (2001)
19. Morrissey, P., Smart, N.P., Warinschi, B.: A Modular Security Analysis of the TLS Handshake Protocol. In: Pieprzyk, J. (ed.) ASIACRYPT 2008. LNCS, vol. 5350, pp. 55–73. Springer, Heidelberg (2008)
20. Niemi, A., Arkko, J., Torvinen, V.: Hypertext Transfer Protocol (HTTP) Digest Authentication Using Authentication and Key Agreement (AKA). RFC 3310 (September 2002)
21. Niemi, V., Nyberg, K.: UMTS Security, 1st edn., ch.8. Wiley Publishing Inc. (2003)
22. OpenID Authentication 2.0 - Final (2010), http://openid.net/specs/openid-authentication-2_0.html (last accessed March 30, 2011)
23. Recordon, D., Reed, D.: OpenID 2.0: a platform for user-centric identity management. In: Proceedings of the Second ACM Workshop on Digital Identity Management, pp. 11–16. ACM (2006)
24. Sovis, P., Kohlar, F., Schwenk, J.: Security Analysis of OpenID. In: Freiling, F.C. (ed.) Sicherheit, GI. LNI, vol. 170, pp. 329–340 (2010)
25. Urueña, M., Busquiel, C.: Analysis of a Privacy Vulnerability in the OpenID Authentication Protocol. In: IEEE Multimedia Communications, Services and Security (MCSS 2010), Krakow, Poland (2010)
26. Windley, P.J.: Digital Identity - Ebook edition. O'Reilly Media (2008)
27. Zhang, J., Warkentin, P., Sankhla, V.: AVISPA model for EAP: Extensible Authentication Protocol, http://www.avispa-project.org/library/EAP_AKA.html (last accessed March 30, 2011)

Can a Mobile Cloud Be More Trustworthy than a Traditional Cloud?

Mufajjul Ali

Orange Lab UK
Building 10, Chiswick Park,
Chiswick, London, W4 5SX
Mufajjul.ali@orange-ftgroup.com

Abstract. Cloud computing is deemed to be the next big trend nebulous. Various sectors have expressed interest in its adoption, including banking, the government, education, manufacturing and telecommunication. With the promise of cost saving and flexibility also comes the greater challenge of security in-particularly "trust". One of the common questions asked by many users is "Can the cloud be trusted?" Telecommunication service providers have been trusted for many years, and have been adopted my millions of users world wide. With the emerging vision of new mobile cloud providers, the ultimate question lies in asking, can a mobile cloud provider be a more trustworthy provider than the traditional ones?

Keywords: Cloud computing, Trust, Security, Telco.

1   Introduction

Cloud computing (CC) is built on many existing tools and technologies reducing the cost of service delivery whilst increasing the speed and agility of service deployment [1]. The core technology behind cloud computing is virtualization; it empowers the whole cloud computing paradigm. The virtualization technology allows the separation of physical hardware and the operating system by creating an abstract layer between both. This allows a greater degree of flexibility by being able to share the same physical resources virtually by more than one OS.

1.1   Cloud Service Models

Cloud computing has various service models to cater for the need of different market segments. Infrastructure as a Service (IaaS) is designed to meet the requirement for businesses that need their infrastructure to be hosted remotely. There are several benefits to this approach; upgrade, management and security of physical resources are provided by the host. The infrastructure can scale at will, and does not require any in-house experts. The down side of this approach is that the logical and physical security is outside the businesses' physical boundaries [2]. Sensitive and private data being hosted remotely raises security concerns and requires trust of the provider.

R. Prasad et al. (Eds.): MOBISEC 2011, LNICST 94, pp. 125–135, 2012.

Platform as a Service (PaaS) on the other hand is designed for companies who require a development environment for developing application/services and hosting facilities remotely. It is especially designed to take away the complexity of setting-up the environment, maintaining and managing software updates.

A relatively new service model proposed by Ali. M [3] is Network as a Service (NaaS). The concept is based on allowing application(s) to dynamically fluctuate the required bandwidth at close to real-time. This greatly improves the performance of bandwidth hungry applications. The customer is charged based on their bandwidth usage, rather than a fixed monthly term.

And finally Software as a Service (SaaS) model is where application(s) and services are hosted by a service provider or a vendor. These application/services are typically accessed remotely via the Internet using a Web browser. From an end user's prospective, this is a fairly attractive model, since installation, upgrade and patches are not required to be managed by the user. User's can be assured that they are using the latest version of an application and are only paying for the actual usage of the service [6].

1.2   Cloud Computing Deployment Models

More recently, there have been four different deployment models defined by the cloud community: private cloud, public cloud, community cloud and hybrid cloud.

Private clouds require the complete cloud infrastructure to be hosted locally with the company's local network. No external network traffic has access to this cloud. There are several benefits to this approach: first, to maximize and optimize in-house resources [7, 8]. Second, the organization has full control of the resources, and finally it is fully secure and operations within the private cloud can be fully trusted.

Public clouds on the other hand are the opposite of private clouds. The complete infrastructure is hosted by a 3rd party provider remotely. The cloud provider has full ownership of the cloud; it may have its own pricing, security and other policies. A public cloud consumer must have significant trust in the provider, since all the data resides under their control.

The community cloud enables several organizations to share the same resources, infrastructure and policies. This collaboration allows them to be more cost effective with better management of available resources.

And finally, a hybrid cloud can be a mixture of different clouds, typically private and public clouds combined together. An organization may use their private cloud for developing in-house service, then migrate to a public cloud for their end users.

2   Background

Over the last 30 years there has been a great advancement in mobile telecommunication networks. The first generation of network, also known as 1G was released in 1980's. It was primarily designed for human to human communication (voice). The advancement from circuit-switch to packet switch network arrived in early 90's. The 2G (GSM) network was designed for machine to human communication (SMS), packet data was added later via the GPRS protocol.

The 3G network, which is hybrid between packet switch and circuit switch, was designed in mind for machine to machine communication. Apart from voice and SMS, it provides various value added services, such as networking games, web browser, etc.

The all IP vision of the 4G network has set the trend for Telcos broadening their horizon in embarking into the service market; this has lead to the optimism of becoming the mobile cloud provider.

2.1   Mobile Cloud Computing (MCC)

Telco operator Orange has been actively involved in defining their concept of mobile cloud computing. Their provisional definition is as follows:

> "Mobile cloud computing is a device-centric cloud that aids the creation, composition and provision of mobile cloud services"

There are two aspects to this definition, firstly the device-centric cloud, which is designed to handle the physical infrastructure requirements. This provides the backbone required for creating innovative services that may not be possible with current deployment models.

Secondly, mobile cloud services are empowered by platforms that provide the necessary building components for the creation, composition and provisioning of mobile cloud services.

The key characteristics of mobile cloud computing is as follows:

State Preservation – capabilities are available for preserving the state and data of an application that can be restored at any point in time (t0, t-1, t-n) on different devices.

Resource Fragmentation – Intelligent resource scheduling (off-loading/on device execution) and optimized algorithms are used to minimize the impact on the device's battery life, RAM, CPU and storage (data sync)

Network Optimization – Edge locality for optimal communication (physical distance) between the device and the service (s) (hosted on the cloud).

Data Sync Management – Near real-time sync of data amongst shared devices for both on line and partially off-line connectivity; this includes appropriate locking and data collision/corruption avoidance management.

Provenance Aware – Advance trust management based on Provenance for each and every operation which are recorded and analyzed to ensure that the operation(s) are legitimate, data confidentiality/integrity is maintained, and accountability/liability is satisfied.

Access Mobility – Device is agnostic to static physical location for access; services can be provisioned and accessed seamlessly.

The characteristics of mobile cloud computing are catered towards meeting device constraints needs (such as increasing processing and memory capability, and device environment portability).

The two service models provided by MCC are PaaS and SaaS (see table 1), which is the same as cloud computing at conceptual level. However, IaaS is not supported by mobile cloud computing - mainly due to security reasons. It however, provides a new service model, known as Network as a Service (NaaS) which was briefly discussed earlier.

Table 1. Comparison of Service models

Type	NaaS	IaaS	PaaS	SaaS
CC	-	X	X	X
MCC	X	-	X	X

The NaaS allows services to be optimized according to the level of bandwidth requirement, which ultimately minimizes network lag and improves the user experience.

Unlike the traditional cloud computing where the client is device agnostic and virtualization mainly powers the server end. With MCC there is also device virtualization; amongst other benefits, this provides smart phone capabilities on featured phones.

The deployment models for MCC is restricted to public and private cloud (see table 2 below)

Table 2. Comparison of deployment model

Type	Public Cloud	Private Cloud	Hybrid Cloud	Community Cloud
CC	-	X	X	X
MCC	X	X	-	-

Many of the core assets of Telco provider Orange (location, presence and others) have been opened up from the core network, and are accessible using restricted API's provided via the MCC.

3   Trust

Trust is a term used in many disciplines such as sociology, psychology, computing and so on. The official Oxford dictionary definition of the term "trust" is as follows: *"Firm belief in the reliability, truth, or ability of someone or something"*

In the notion of computing, trust is orthogonal to security; trusted components/entities are required to build a secure system. In the context of the cloud, three areas concerning trust are (see fig 1 below): Security, Availability and Performance.

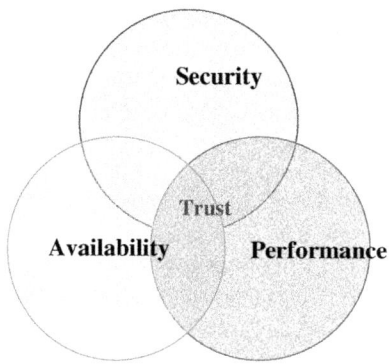

Fig. 1. Notion of trust

These areas are of most concern when it comes to trusting cloud providers and the services they provide. In a recent poll carried out by an IDC enterprise panel [5], 87.5% of voters voted security as their main concern, followed by availability at 83.3%, and performance at 82.9%.

MCC characteristics are *inherently* leverages on the existing trust of Telco network, and it is focused on providing the much needed building blocks that can enhance the customer's service trust.

The characteristics defined in section 2.11 can be complimentary to the "Trust" factor shown above.

The first characteristic ("State Preservation") can improve the usability of services by being able to restore/resume state of the service(s) from any device that are considered to be trusted by the IAM (Identity and management system) mechanism. This is particularly useful in cases where the battery life of the device is at minimal and there is an explicit need for switching of devices; which can greatly improve the performance factor of the service.

Characteristic two ("Resource Fragmentation") on the other hand is an important feature of the MCC. The sole purpose of this is to alleviate the performance and computational issues faced by many devices by extending its capability to almost limitless. The customer would greatly benefit from the enhanced performance of their services. One point to note is that this feature is service specific; the performance factor is heavily dependent on how the service is fragmented and algorithms have been implemented.

Characteristic three ("Network Optimization") can greatly enhance the performance of the service access, by ensuring that the optimal distance of the physical locality of data/service is selected. This would greatly benefit users who are constantly on move, possibly from one country to another or moving between cities. However, this is assuming that latency is constant amongst the nodes, and same level of bandwidth between user's device and nodes.

Characteristic four ("Device Sync Management") is more specific to the device content. It provides the facilities for structured data to be synced between the device and the cloud. The benefits are that a user can trust their data will remain safe even if the device is lost or stolen or, can access his/her data from different devices.

Characteristic five ("Provenance Aware") is probably the most important feature related to trust in MCC. The main purpose of this feature is providing the end user with greater transparency by providing provenance data for each individual application. The provenance data has many usages. Firstly, it can be used for determining the liability and accountability in case any faults occur in the cloud. Secondly, it can be used to detect anomalies, and can also be used for better policy control.

Characteristic six ("Access Mobility") is designed for continuous service access. If a device is on Wifi, and the signal is getting weaker due to mobility, the device would automatically switch to the stronger 3G network, and vice versa. This ensures that minimal disruption is caused to the network access. The advantage of this approach is that it does not explicitly require user's interaction.

3.1   Security

Security is one of the key factors of trust. Many cloud providers are unlikely to guarantee the security of data [10], hence breeching the DPA 1998 [11]. There is an explicit need for security of the data as it faces the potential threats from forth-coming cloud malware. Cloud malware is likely to be a new breed of innovative and sophisticated techniques being developed and used to compromise the cloud.

3.2   Platform

Malicious threats such as viruses, worms and trojans [17] are also major security concerns for companies and organizations. Statistics shows that the Windows platform is most susceptible to malware (see table 3 below). Linux and Mac platforms on the other hand are less prone to the attack; this is may be due having a more robust OS core, or due to occupying a smaller market share compared to Windows.

Table 3. Platfrom threats

OS	Viruses	Worm	Trojan
Windows	60000+[18]	1000+	1000+
Linux	40+	Limited	Limited
Mac OS	5+	Limited	Limited
Symbian	10+ [21]	-	52
Android	1[15]	-	1[16]
Iphone	1 [12]	1 [13]	1 [14]

Mobile OSes however are still relatively safe for terminal operations for MCC, despite new potential malware threats. The current wave of threats on Android and Linux only affect devices that have been hacked ('jail-broken').

3.3   Network/Infrastructure

Traditionally, an IP Telecoms network (3G) is considered to be more secure than the standard Ethernet network (Internet) (see table 4 below comparing a specific Telco Operator's network with the Ethernet network).

Table 4. Network Security

OS	3G Network	Ethernet Network
Authentication	Sim-based	None/application specific
Network access	Secure, Authentication required	Non-secure, ISP specific authentication
Data	Ciphered	Plain/application specific
Network standardization	3GPP standardized	IETF
Network implementation	Operator specific	Shared

There are several reasons for greater Telco security. Firstly, a Telco's core network is standardized by 3GPP. Each of the components is well defined and huge amounts are invested each year on maintenance and upgrades. This is to ensure that there are no defects and the highest level of quality of service is achieved.

Secondly, the physical core network is generally closed from outside; only the operator has access to many of its sensitive assets, such as the HLR. However, with the vision of providing APIs for third-parties to create innovative services, it is slowly becoming more and more open.

Thirdly and most importantly, access to the network requires secure authentication, this is not the case in an IP network. Each device is authorized by using a non-temperable SIM. And most significantly, all communication is carried out under a secure communication channel. The actual transmissions of data are ciphered for additional security [20]. This is contrary to the Internet where prior authentication is not required, as long as there is an ISP providing the Internet connection. A secure IP communication channel is optional (SSL/VPN) and it is based on specific application/service needs.

3.4   Availability

Availability is a key measurement of Quality of Service (QoS). It is defined by the equation below (See figure 1), which calculates the uptime of a service during its life span.

$$A = \frac{E[\text{Uptime}]}{E[\text{Uptime}] + E[\text{Downtime}]}$$

Fig. 2. Availability Equation

Google have recently been sued for inappropriate security on its Cloud services [9] and a recent problem with their mail system caused dismay amongst many businesses and consumers, who were denied service for several hours. This further highlights the potential danger of fully trusting cloud service providers.

Table 5. Network Avialability

Feature	Availability
CC	99.5 [4]
MCC	99.9999

Telcos however are not known for unexpected denial of critical services. They have very high threshold values for availability (see table 5 above) for their PSTN service. Given the fact that operator such as Orange owns their own broadband network. It should be possible to offer very high availability for end-to-end services. The high availability offered by Amazon is server end only. The true end-to-end availability would be dependent on the ISP providing the connection, which may significantly affect the overall availability threshold. The superior end-to-end availability promise of MCC would be the preferred choice of being highly trusted by critical sectors such as Banking, Medical and the Government.

3.5   Performance

Performance is determined by systems or applications performing to levels either defined by a contractual obligation or industry-recognized acceptable levels. It is directly related to the second factor 'availability' where the higher availability is likely to be paralleled by high performance levels.

There are various matrices can be used to determine the performance level. MCC, and CC both rely on physical networks to deliver services to the customers; which makes network delivery capability a good matrix to measure trust in performance.

Table 6. Comparison of deployment model

Type	3G	4G	WiMax	Broadband
Speed	3.6-7.2Mb/s	>100 Mb/s [19]	108 Mb/s	50 Mb/s
Coverage	93% UK	Yet to be released	50km (fixed), 5-15 km (mobile)	Physical connection
Latency	Increases in built up areas,	Not yet known	Increases with distance	Application specific
Speed	3.6-7.2Mb/s	>100 Mb/s [19]	108 Mb/s	50 Mb/s

The speeds defined in Table 6 above are theoretically achievable speeds in UK; the actual speeds may be slightly lower due to overall network overhead and latency.

The speed of a 3G network is fairly modest; it is likely to perform reasonably well with applications and services that require low bandwidth, but may not be optimized for bandwidth-hungry applications. However, with the proposal of 4G this obstacle can be minimized. The speed of broadband is a highly attractive proposition for performance for CC, but would not have the benefit of "anywhere anytime" capability of MCC and does not have the support of NaaS model.

3.6   Other Trust Factors

Reputation - With any cloud computing service, it's important that the provider have a trusted relationship with those people using the service based on reputation"[23] also highlights a significant point that could sway the trust factor in favour of MCC.

Large Telco providers have been serving customers for several decades, and within this period, they have served and built a trusted relationship with millions of their customers. As general consensus suggests, that based on reputation, customers are more likely to trust their personal data with a Telco provider than traditional cloud providers such as Google.

It is not common for Telcos to loose their customers' data, although recently a Telco provider had their repetition slightly dented by loss of their customers' personal data on their 'Sidekick' devices [22]. The device stored its data in the cloud (provided by Microsoft) with the mishap affecting thousands of its customers.

Community/Experts - It is commonly accepted that the opinions of experts are trusted over those of lay-persons. In this regards, Telcos follow the principle of standardization followed by implementation and MCC is no stranger to this approach. CC on the other hand follows the 'implementation that may lead to standardization' approach. This may carry a greater risk of mistrust as unproven services may be provided to customers, possibly containing unknown defects.

4   Conclusion

Three areas of concerns have been highlighted in this paper; Security, Availability and Performance, which mainly contribute to level of trust by service providers. Mobile cloud models seem to provide more robust security mechanisms than traditional cloud computing models at the infrastructure level. However MCC is lagging behind when it comes to delivery of service to the client. The expected arrival of 4G networks could be the holy-grail for realizing MCC's service performance potential and making the NaaS a reality.

Cloud computing is certainly a more mature technology than Mobile cloud computing, and in the short term customers may be more willing to trust it. However, given the nature of malware threats that exist on CC, MCC would be a better alternative.

Reputation is also an important indicator for trusting a provider. Telco's reliable service history will have a positive affect on gaining trust in MCC solutions, but this reputation itself may not be enough for customers to fully trust in MCC to give up control of their sensitive and person data. Alongside reputation, transparency will also be paramount. The customer should see be able to access the full history of every action, transaction, operation occurred on their data, on request. At present, this is not guaranteed; however, with the emergence of provenance technology, it might be sooner rather than later to become a reality.

References

[1] Voas, J., et al.: Cloud computing: New wine or just a new bottle? vol. 11(2), pp. 15–17. IEEE Computer Society (March/April 2009)
[2] John, W., et al.: Cloud computing: Implementation, management and security, pp. 153–180. CRC Press (2010)
[3] Ali, M.: Green Cloud on the Horizon. In: Jaatun, M.G., Zhao, G., Rong, C. (eds.) CloudCom 2009. LNCS, vol. 5931, pp. 451–459. Springer, Heidelberg (2009)
[4] Amazon (2010), http://aws.amazon.com/ec2/
[5] http://blogs.idc.com/ie/wp-content/uploads/2009/12/idc_cloud_challenges_2009.jpg
[6] Khalid, A.: Cloud computing: Applying Issues in Small. In: 2010 International Conference on Signal Acquisition and Processing. IEEE (2010)
[7] Dillon, T., et al.: Cloud computing: Issues and Challenges. In: 24th International Conference on Advanced Information Networking and Applications. IEEE (2010)
[8] Jesen, M., et al.: On Technical Security Issues in Cloud Computing. In: IEEE International Conference on Cloud Computing (2009)
[9] Marshall, R.: Google explains Gmail troubles (February 2009), http://www.vnunet.com/vnunet/news/2237201/google-gmail-troubles-explained
[10] Mari, A.: Cloud computing could bring security threats (February 2009), http://www.vnunet.com/computing/news/2237013/cloud-computing-bring-security
[11] Data Protection Act (1998), http://www.opsi.gov.uk/Acts/Acts1998/ukpga_19980029_en_1,1998
[12] Claire, B.: First malicious virus hit iPhone...but only on mobiles cracked by users (November 2009), http://www.dailymail.co.uk/sciencetech/article-1230223/First-malicious-virus-hits-iPhone-mobiles-cracked-users.html
[13] BBC news, Worm attack bites at Apple iPhone, http://news.bbc.co.uk/2/hi/technology/8349905.stm
[14] Dinah, G.: Trojan targets iPhone users (August 2010), http://www.talktalk.co.uk/technology/news/articles/trojan-targets-iphone-users.html
[15] Mills, E.: First SMS sending Android Trojan reported (August 2010), http://news.cnet.com/8301-27080_3-20013222-245.html
[16] Beavis, G.: Expensive virus hit Android Users, fake application sucks mega-bucks from your phone (August 2010), http://www.techradar.com/news/phone-and-communications/mobile-phones/expensive-virus-hits-android-users-709260
[17] Ali, M., et al.: A Review of Security Threats on Smart Phones. In: ICGeS (February 2005), Conference Paper (April 2005)

[18] The register, Linux vs windows viruses (October 2003),
     http://www.theregister.co.uk/2003/10/06/
     linux_vs_windows_viruses/
[19] Santhi, K.R., et al.: Goals of True Broad band's Wireless Next Wave (4G-5G). In: 2003
     IEEE 58th Vehicular Technology Conference, VTC 2003-Fall, vol. 4, pp. 2317–2321
     (May 2004)
[20] Smith, C., et al.: 3G Wireless Networks. McGrow-Hill Telecom professional (2002)
[21] Top 10 Most Dangerious Viruses on Symbian based Cell phones (April 2010),
     http://www.worldinterestingfacts.com/lifestyle/
     top-10-most-dangerous-cell-phone-viruses-on-symbian-based-
     cell-phone.html
[22] Malik, When the Cloud Fails: T-Mobile, Microsoft Lose Sidekick Customer Data
     (October 2009), http://gigaom.com/2009/10/10/when-cloud-fails-t-
     mobile-microsoft-lose-sidekick-customer-data/
[23] BBC (April 2010), http://news.bbc.co.uk/1/hi/programmes/click_
     online/8625625.stm

Fingerprint Recognition
with Embedded Cameras on Mobile Phones

Mohammad Omar Derawi, Bian Yang, and Christoph Busch

Norwegian Information Security Laboratory, Gjvik University College, Norway
{mohammad.derawi,bian.yang,christoph.busch}@hig.no
http://www.nislab.no/

Abstract. Mobile phones with a camera function are capable of capturing image and processing tasks. Fingerprint recognition has been used in many different applications where high security is required. A first step towards a novel biometric authentication approach applying cell phone cameras capturing fingerprint images as biometric traits is proposed. The proposed method is evaluated using 1320 fingerprint images from each embedded capturing device. Fingerprints are collected by a Nokia N95 and a HTC Desire. The overall results of this approach show a biometric performance with an Equal Error Rate (EER) of 4.5% by applying a commercial extractor/comparator and without any preproccesing on the images.

Keywords: biometric systems, fingerprint recognition, mobile phone cameras, user authentication.

1 Introduction

Current mobile devices implement various new kinds of applications such as taking photos, and movie shooting by using embedded camera devices. This progress was made possible by the evolution of miniaturized embedded camera technology. Mobile devices – particularly mobile phones – are being found in almost everyone's hip pocket these days all over the world. Almost all newer cell phones now-a-days have embedded camera devices, and some of those have more than over 5 mega-pixel image cameras.

From a security point of view, the issues related to ever-present mobile devices are becoming critical, since the stored information in them (names, addresses, messages, pictures and future plans stored in a user calendar) has a significant personal value. Moreover, the services which can be accessed via mobile devices (e.g., m-banking and m-commerce, e-mails etc.) represent a major value. Therefore, the danger of a mobile device ending up in the wrong hands presents a serious threat to information security and user privacy. According to the latest research from Halifax Home Insurance claims, 390 million British pounds a year is lost in Britain due to the theft of mobile phones. With the average handset costing more than 100 British pounds, it is perhaps not surprising that there are more than 2 million stolen in the UK every year [1].

R. Prasad et al. (Eds.): MOBISEC 2011, LNICST 94, pp. 136–147, 2012.

Authentication is an area which has grown over the last decades, and will continue to grow in the future. It is used in many places today and being authenticated has become a daily habit for most people. Examples of this are PIN code to your banking card, password to get access to a computer and passport used at border control. We identify friends and family by their face, voice, how they walk, etc. As we realize there are different ways in which a user can be authenticated, but all these methods can be categorized into one of three classes [2]. The first is *something you know* (e.g., a password), the second is *something you have* (e.g., a token) and the third is *something you are* (e.g., a biometric property).

Unlike passwords, PINs, tokens etc. biometric characteristics cannot be stolen or forgotten. The use of biometric was first known in the 14th century in China where "Chinese merchants were stamping childrens palm- and foot prints on paper with ink in order to distinguish young children from one another". Approximately after 500 years has passed, the first fingerprinting was used for identification of persons. In 1892, the Argentineans developed an identification system when a woman was found guilty of a murder after the investigation police proved that the blood of the womans finger on the crime scene was hers. The main advantage of biometric authentication is that it establishes an explicit link to the identity because biometrics use human *biological* and *behavioral* characteristics. The first mentioned are the biometrics derived directly from the part of a human body. The most used and prominent examples are the fingerprint, face, iris and hand recognition. The behavioral characteristics are the biometrics by persons behavioral characteristics, such as gait-recognition, keystroke recognition, speech/voice recognition and etc.

Many fingerprint recognition algorithms perform well on databases that had been collected with high-resolution cameras and in highly controlled situations [3]. Recent publications show that the performance of a baseline system deteriorates from Equal Error Rate (EER) around 0.02 % with very high quality images to EER = 25 % due to low qualities images [4]. Thus active research is still going on to improve the recognition performance. In applications such as fingerprint authentication using cameras in cell phones and PDAs, the cameras may introduce image distortions (e.g., because of fish-eye lenses), and fingerprint images may exhibit a wide range of illumination conditions, as well as scale and pose variations. An important question is which of the fingerprint authentication algorithms will work well with fingerprint images produced by cell phone cameras?

However, recent research [5,6] have shown that by using low-cost webcam devices it is possible to extract fingerprint information when applying different pre-processing and image enhancements approaches. In this paper we present fingerprint recognition as means of verifying the identity of the user of a mobile phone. The main purpose of this paper is to study how it is possible to lower down the user effort while keeping the error rates in an acceptable and practical range. Therefore, this proposal is a realistic approach to be implemented in mobile devices for user authentication. To address this issue, we collected a fingerprint database at the Norwegian Information Security Laboratory using two different cell phone cameras, namely the Nokia N95 and HTC Desire where details mentioned later.

2   Fingerprint Recognition

Fingerprint recognition is the most matured approach among all the biometric techniques ever discovered. With its success of use in different applications, it is today used in many access controls applications as each individual has an immutable, unique fingerprint. The hand skin or the finger skin consists of the so called friction ridges with pores. The ridges are already created in the ninth week of an individuals fetal development life [7], and remains the same all life long, only growing up to adult size, but if severe injuries occur the skin may be reconstructed the same as before. Researchers have found out that identical twins have fingerprints that are quite different and that in the forensic community it is believed that no two people have the same fingerprint [8].

Many capture device technologies have been developed over the last decades replacing the old ink imaging process. The old process was based on sensing ridges on an individuals finger with ink, where newer technologies uses a scanner placing the surface of the finger onto this device. Such technologies are referred to as live-scan and based on four techniques [9]:

Frustrated total internal reflection (FTR) and optical methods is   a
first live scan technology. Figure 1 illustrates, how the reflected signal is acquired by a camera from the underside of a prism when a finger touches the top of the prism. The typical image acquisition surface of 1 inch by 1 inch is converted to 500 dots per inch (DPI) using either charge coupled device (CCD) or complementary metal oxide semiconductor (CMOS) camera.

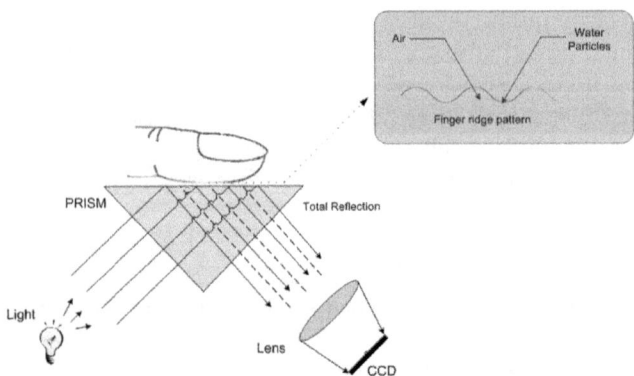

Fig. 1. Optical fingerprint sensing by frustrated total internal reflection

CMOS Capacitance. The ridges and valleys create different charge accumulations, when a finger hits a CMOS chip grid. This charge is converted to an intensity value of a pixel using various competing techniques such as alternating current (AC), direct current (DC) and radio frequency (RF). The typical image acquisition surface of 0.5 inch by 0.5 inch is converted to 500 dots per inch (DPI). The resultant images also have a propensity to be affected by the skin dryness and wetness.

Ultrasound Sensing. The thermal sensor is developed by using pyro-electric material, which measures temperature changes due to the ridge-valley structure as the finger is swiped over the scanner and produces an image. In this case the skin is a better thermal conductor than air and thus contact with the ridges causes a noticeable temperature drop on a heated surface. This technology is claimed to overcome the dryness and wetness of the skin issues of optical scanners. But the resulting images are not affluent in gray value images. The thermal sensor is becoming more popular today, because they are small and of low cost. Swipe sensors based on optical and CMOS technology are also available as commercial products.

3   Data Collection

3.1   Rationale

Besides fingerprint recognition systems deployed for applications with high-security requirements such as border control [10,11] and forensics [12], fingerprint recognition is supposed to be promising for consumer markets as well for many years [13,14]. In the meanwhile, privacy concerns over fingerprint recognition technologies' deployment in non-high-security applications have been raised [15,16] and thus leads to a refrained development of biometrics in consumer market in recent years compared with the rapid development in the public sectors such as border control, critical infrastructure's access control, and crime investigations.

We suppose there are at least two ways to alleviating these privacy concerns. Biometric template protection [17,18] is one of the most promising solutions to provide a positive-sum of both performance and privacy for biometric systems' users. The European Research Project TURBINE [19] demonstrated a good result in both performance and privacy of the ISO fingerprint minutiae template based privacy-enhancement biometric solutions. On the other hand, for the consumer market, we think using customers' own biometric sensors will also help alleviate the customers' privacy concerns. That is the motivation of this paper to try using cell phone cameras as sensors for fingerprint sample collection.

Obviously, for applications requiring high security, subjects' own biometric sensors may not be suitable for data collection unless the cell phone can be authenticated as a registered and un-tampered device in both software and hardware aspects, which is difficult to realize for a normal consumer electronics that is out of the control of the inspection party. However for consumer market, cell phone can be deemed nowadays as a secure device accepted by many customers, e.g, many banking services send transaction password, TAN code or PIN code via SMS to customers' cell phone. So in this paper we assume biometric data collection by the customers' cell phone cameras will not raise more privacy and security concerns to the customers than the cell phone based banking services.

In the meanwhile we expect technical challenges in quality control to the cell phone camera captured samples, especially from the sample image processing aspects such as bias lighting conditions and unstable sample collection environment caused by hand-holding. In addition, most existing cell phone cameras are not designed for biometric use and accurate focusing will always be a challenge for fingerprint image capturing. We address these potential challenges in this paper in a simplified way to investigate whether cell phone camera can generate good quality samples and corresponding good biometric performance in a relative stable data collection environment.

3.2 Data Collection Steps

As there is no standard benchmark database available for fingerprint images captured by digital camera, we constructed an independent database. The image database is comprised of 22 subjects from which fingerprint images were taken with a cell phone camera. The fingerprint data used in this paper are captured by two commercial sensors as shown in Figure 2. The cell cameras used were

Fig. 2. Left: CMOS Sesnor (HTC Desire), Right: CMOS Sensor (Nokia N90) and a cropped/contrasted fingerprint image from each cell, at the same scale factor

Carl Zeiss Optics from Nokia N95 and HTC Desires' embedded camera. Further detailed information of the sensors is described in Table 1.

The constructed independent database comprises of 1320 fingerprint images. These images stem from 220 finger instances, where each instance was captured 6 times. The images are stored in the internal memory of the phones and all the

Table 1. Cell phone camera setting for fingerprint image acquisition

Cell Phone	Nokia N95	HTC Desire
Lens Type	CMOS, Tessar lens	CMOS
Mega Pixel	5.0	5.0
Resolution	2592x1944	2592x1552
Flash	LED Flash	LED Flash
ISO Speed	100 - 800	52
Auto-Focus	Yes	Yes

Fig. 3. Setup for the Nokia N95 capture device

images were collected in the cameras "Burst Mode". For evaluating the performance of various algorithms under different settings, the Nokia N95 was fixed placed on a hanger as illustrated in Figure 3 where images were taken by a human operator holding the phone and capturing images for the HTC Desire. The image capture was performed inside a laboratory with normal lighting conditions.

4   Evaluation

As can be seen in Figure 4, the user initially presents its biometric characteristic (i.e., capturing the fingerprint) to the sensor equipment (i.e. camera in a mobile phone), which captures it as captured biometric sample. After preprocessing this captured sample, features will be extracted from the sample. In case of fingerprint biometrics, these features would typically be minutia points. The extracted features can then be used for comparison against corresponding features stored in a database, based on the claimed identity of the user. The result of the comparison is called the *similarity score* S, where a low value of S indicates little similarity, while a high value indicates high similarity. The last step is to compare the similarity score S to a predefined system *threshold* T, and output a decision based on both values. In case the similarity score is above the threshold $(S > T)$ then the user is accepted as genuine, while a similarity score below the threshold $(S < T)$ indicates an impostor who is rejected by the

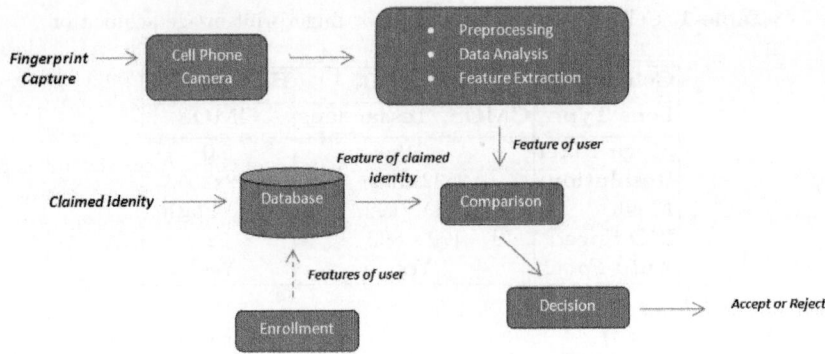

Fig. 4. A traditional verification process

system. Obviously the biometric features of the user must initially be stored in the database before any comparison of a probe feature vector can take place. This is done during the *enrolment phase*. During the enrolment biometric samples are captured from the biometric characteristic, after which it is processed and features are extracted. The extracted data is now stored in a database and linked to the identity of the user who enrolled. The stored data in the database is referred to as the *reference template* of the user. In case of fingerprint biometrics it is a common approach to derive the features from multiple captured samples and generate a single minutiae template.

4.1   Feature Extraction

In order to measure the sensor performance we have applied the Neurotechnology, Verifinger 6.0 Extended SDK commercial minutia extractor for the feature extraction. The SDK includes functionality to extract a set of minutiae data from an individual fingerprint image and to compute a comparison-score by comparing one set of minutiae data with another. Both SDKs support open and interoperable systems as the generated minutiae templates can be stored according to the ISO or ANSI interchange standard.

4.2   Feature Comparison

We compared the verification results of the Neurotechnology algorithm on the processed images. For each algorithm the error rates were determined based on a threshold separating genuine and impostor scores. The False Match Rate (FMR) and False None-Match Rate (FNMR) were calculated. The calculation of FMR and FNMR is done in the following way. We have collected N data samples from each of M participants, then we have calculated similarity scores between two samples, either stemming from one finger instance or from two different instances. A similarity score between two samples from the same source is called a genuine score, while an impostor score is the similarity score between two samples from different instances. Given our setting, we can have $N * M$ data samples from which we can calculate the total number of $N_{Gen} = \frac{M*N*(N-1)}{2}$ different genuine scores and $N_{Imp} = \frac{M*N*(M-1)*N}{2}$. Given these sets of genuine and impostor scores we can calculate FMR and FNMR for any given threshold T as follows:

$$FMR(T) = \frac{Number\ of\ incorrectly\ accepted\ impostor\ images \geq T}{Total\ number\ of\ impostor\ images} \quad (1)$$

$$FNMR(T) = \frac{Number\ of\ incorrectly\ rejected\ genuine\ images < T}{Total\ number\ of\ genuine\ images} \quad (2)$$

From this, we can find the point where FNMR equals FMR, or in other words the Equal Error Rate (EER). This rate is very common used value which is being used to compare different systems against each other, and it roughly gives an idea of how well a system performs.

The images that were generated with the mobile phones encode the finger position according to Table 2 and the equal error rates retrieved corresponding to the finger codes are overviewed in Table 3

Table 2. Finger position codes according to ISO 19794-2

Finger Position	Code
Right thumb	1
Right index finger	2
Right middle finger	3
Right ring finger	4
Right little finger	5
Left thumb	6
Left index finger	7
Left middle finger	8
Left ring finger	9
Left little finger	10

Table 3. EERs of cell phone fingerprint recognition. Numbers are in percentage

Cell Phone	1	2	3	4	5	6	7	8	9	10	all
Nokia N95:	5.77	5.92	5.11	7.36	5.43	2.98	0.0	0.43	6.26	5.45	4.66
HTC Desire:	11.73	11.43	23.62	21.17	16.01	10.98	8.47	15.37	16.11	15.96	14.65

In general we see that the left index finger (code 7) has performed best for both phones with EER of 0.0% and 8.47%. The overall performance (cross comparison of all ten fingers) which can be seen in column *all* for Nokia N95 performs significantly better than the Desire. This is so because of various reasons. The Nokia was placed in fixed way on the holder while capturing. Furthermore, the Nokia was set to an internal close-up mode setting. This mode is ideal for capturing details of small objects within a distance between 10 and 60 cm. Here we had to ensure that the auto-focus always resulted in better quality images at a small distance when capturing the fingerprints, whereas the HTC was manually adjusted by the human operator. Thus, this means that the Nokia N95's auto-focus was performing slightly better than the HTC Desire.

5   Discussion

Since personal mobile devices at present time only offer means for explicit user authentication, this authentication usually takes place one time; only when the mobile device has been switched on. After that the device will function for a long time without shielding user privacy. As of today the majority of Internet users are expecting a transparent transition of services from the wired to the wireless mobile world. As personal mobile devices such as Apple's iPhone, T-Mobile's G1 or Nokia's S60 become more popular the ordinary user is expecting and using the full range of Internet services in the mobile Internet, since former limitations with regard to screen size and interaction capabilities (zooming, "copy and paste" functionality etc.) disappeared recently. In fact many users are even extending their expectations from their home and office environment, as they enjoy typical mobile features, such as location-based services, which are supported by widespread GPS-features.

On the contrary users tend to ignore the risks, which they accept while operating Internet services from their mobile device. Not only sensitive information is accessible from the mobile device but also transactions on the stock market and other critical services, which grant access to financial assets. At the same time mobile devices are more exposed to the public and thus there is likelihood that a mobile device is lost or stolen in an unattended moment. This threat is shown by the number of approx. 10.000 mobile phones, which were left in London taxis every month in 2008 [20].

It is obvious that a mobile Internet can only exist, if there is a strong link between the mobile device and the authorized user of that specific device. This requires that proper access control mechanisms are in place, to control that the registered user and only the registered user operates the mobile device. Unfortunately most mobile devices are operated today with knowledge-based access control only, which is widely deactivated due to the associated inconvenience.

A promising way out of these pressing problems is to implement on mobile devices secure biometric access control mechanisms, which provide a non-reputable approach based on the observation of biological characteristics (i.e. the fingerprint) of the registered user. The aim of a biometric access control process is, to determine whether the biometric characteristic of the interacting subject and the previously recorded representation in the reference data match.

A possible application scenario of a the fingerprint biometric user verification system in a mobile device could be as follows; When a device such as a mobile phone, is first taken into use it would enter a "practicing" learning mode where the high quality fingerprints data are processed and stored. Password-based or PIN code user authentication would be used during the learning session. If the solidity fingerprint biometrics was sufficient enough, the system would go into a biometric authentication "state", a state that will need confirmation from the owner. In this state the system would asynchronously verify the owner's identity every time the owner wanted to authenticate.

6   Conclusion

The cell phone camera database has been used to study the performance of some fingerprint verification algorithms in a first step towards real-life situations. The database has scaled and posed distortions in addition to illumination. The camera lens' cause further distortion in the images with changes in orientation.

The novel biometric method for frequent authentication of users of mobile devices proposed in this paper was investigated in a technology test. It contained fingerprints data. The recognition resulted in different performances of using one minutia extractor and comparator. The best algorithm performance gained resulted in an EER of 4.66.% for the Nokia N95. Looking forward into which finger was performing best, then we observe an EER of 0.0% for the left index finger as well.

The shown results suggest the possibility of using the proposed method for protecting personal devices such as PDAs, smart suitcases, mobile phones etc. In a future of truly pervasive computing, when small and inexpensive hardware can be embedded in various objects, this method could also be used for protecting valuable personal items. Moreover, reliably authenticated mobile devices may also serve as an automated authentication in relation to other systems such as access control system or automated external system logon.

Acknowledgments. We would like to thank Simon McCallum and Jayson David Mackie for using their cell phones for the experiment. Furthermore, we would like to thank all volunteers participating in the data collection.

References

1. Mobile phone theft increasing across the uk,
   http://www.insure4u.info/home-insurance-mobile/
   mobile-phone-theft-increasing-across-the-uk.html
   (accessed March 30, 2011)
2. Shah, S.U., Fazl e Hadi, Minhas, A.A.: New factor of authentication: Something you process. In: International Conference on Future Computer and Communication, pp. 102–106 (2009)
3. Nist image group's fingerprint research,
   http://www.itl.nist.gov/iad/894.03/fing/fing.html (accessed February 13, 2011)
4. Gafurov, D., Bours, P., Yang, B., Busch, C.: Guc100 multi-scanner fingerprint database for in-house (semi-public) performance and interoperability evaluation. In: International Conference on Computational Science and its Applications, pp. 303–306 (2010)
5. Mueller, R., Sanchez-Reillo, R.: An approach to biometric identity management using low cost equipment. In: International Conference on Intelligent Information Hiding and Multimedia Signal Processing, pp. 1096–1100 (2009)
6. Hiew, B.Y., Teoh, A.B.J., Yin, O.S.: A secure digital camera based fingerprint verification system. J. Vis. Comun. Image Represent. 21, 219–231 (2010)
7. Fetal development, http://www.pregnancy.org/fetaldevelopment (accessed February 13, 2011)
8. Pankanti, S., Prabhakar, S., Jain, A.K.: On the individuality of fingerprints. IEEE Trans. Pattern Anal. Mach. Intell. 24, 1010–1025 (2002)
9. Bolle, R., Connell, J., Pankanti, S., Ratha, N., Senior, A.: Guide to Biometrics. Springer (2003)
10. Safety & security of u.s. borders: Biometrics,
    http://travel.state.gov/visa/immigrants/info/info_1336.html   (accessed February 13, 2011)
11. 2004/512/ec: Council decision of 8 june 2004 establishing the visa information system (vis), http://eur-lex.europa.eu/LexUriServ/LexUriServ.do?uri=CELEX:32004D0512:EN:NOT (accessed February 13, 2011)
12. Method for fingerprint identification,
    http://www.interpol.int/public/Forensic/fingerprints/
    WorkingParties/IEEGFI/ieegfi.asp (accessed February 13, 2011)
13. Fingerprint payment heads for the uk,
    http://www.talkingretail.com/news/industry-news/
    fingerprint-payment-heads-for-the-uk (accessed February 13, 2011)
14. Fingerprint biometric retail pos systems on show,
    http://www.prosecurityzone.com/News/Biometrics/Fingerprint
    _recognition/Fingerprint_biometric_retail
    _pos_systems_on_show_16250.asp#axzz1Dr8cNsOu (accessed February 13, 2011)
15. Retailers fingerprint plans prompt privacy concerns,
    http://www.computing.co.uk/ctg/news/1826631/
    retailers-fingerprint-plans-prompt-privacy-concerns
    (accessed February 13, 2011)

16. Europe tells britain to justify itself over fingerprinting children in schools, `http://www.telegraph.co.uk/news/worldnews/europe/eu/8202076/Europe-tells-Britain-to-justify-itself-over-fingerprinting-children-in-schools.html` (accessed February 13, 2011)
17. Jain, A.K., Nandakumar, K., Nagar, A.: Biometric template security. EURASIP J. Adv. Signal Process, 113:1–113:17 (January 2008)
18. ISO/IEC FDIS 24745 Information technology – Security techniques – Biometric information protection, FDIS (February 2011)
19. Eu fp7 integrated project - trusted revocable biometric identities, `http://www.turbine-project.org/` (accessed February 13, 2011)
20. Security park: 60,000 mobile phones left in london taxis in the last six months, `http://www.securitypark.co.uk/security_article262056.html` (accessed February 13, 2011)

Policy Driven Remote Attestation

Anandha Gopalan, Vaibhav Gowadia, Enrico Scalavino, and Emil Lupu

Department of Computing
Imperial College London
180, Queen's Gate
London, SW7 2RH, U.K.
{a.gopalan,v.gowadia,e.scalavino,e.c.lupu}@imperial.ac.uk

Abstract. Increasingly organisations need to exchange and share data amongst their employees as well as with other organisations. This data is often sensitive and/or confidential, and access to it needs to be protected. Architectures to protect disseminated data have been proposed earlier, but absence of a trusted enforcement point on the end-user machine undermines the system security. The reason being, that an adversary can modify critical software components. In this paper, we present a policy-driven approach that allows us to prove the integrity of a system and which decouples authorisation logic from remote attestation.

Keywords: Remote Attestation, Trusted Platform Module, Policy based attestation.

1 Introduction

Governments, businesses and social organisations require the exchange of data between employees as well as with other organisations. This data is often sensitive and/or confidential but its exchange is vital for the successful functioning of these organisations. In particular, this becomes more important with more and more organisations and employees using mobile devices which increases the probability of the data falling into the wrong hands due the thefts of these devices.

Privacy and business confidentiality requirements demand that only authorised people should be granted access to sensitive data, and the usage of sensitive data needs to be controlled even after the data has been disseminated to data consumers. Data may reside at many locations such as server-side data stores, end-user machines, and portable disks. Controlling usage of sensitive data irrespective of its location requires that the sensitive data is always encrypted when stored or transmitted to force all access through a trusted Policy Enforcement Point (PEP). Only the PEP should be able to obtain decryption keys when permitted by a trusted Policy Decision Point (PDP).

Sandhu et al. [14,18] and Gowadia et al. [7] have described various security architectures to protect disseminated or shared data. A common factor among these architectures is the need for a trusted policy enforcement point. A potential threat is that an adversary can modify critical software components such

R. Prasad et al. (Eds.): MOBISEC 2011, LNICST 94, pp. 148–159, 2012.

as enforcement and policy evaluation components, or the underlying operating system. Such attacks can be mitigated in corporate environments by restricting the rights of users, so that they are unable to modify critical files on a system where sensitive data may be used. However, this assumption does not hold when dealing with a malicious insider, when the equipment is stolen or otherwise falls into the wrong hands. This is especially true in the case of using mobile devices due the high risk of theft. It is then necessary to increase the level of assurance provided. To provide a high assurance of system integrity we must rely on a trusted hardware component (e.g., the Trusted Platform Module (TPM), whose specifications are defined by the Trusted Computing Group (TCG) [23]). The TCG is working on the Mobile Trusted Module (v2.0) specification to improve mobile phone security.

The process of verifying integrity of remote systems using TPMs is called *remote attestation*. Remote attestation requires that a trusted verifier is able to verify integrity based on a digitally signed list of checksums provided by the data consumer. The process of verifying this evidence requires the verifier to maintain a large up to date database of acceptable software components that may exist on a data consumer's system. It is often desired that such a task be outsourced to a specialist. Managing trusted reference data for computing software measurements is a cumbersome task and by outsourcing the task to a specialist, we can reuse work done by the specialist verifier, thus reducing the cost of maintaining updates. However, out-of-the-box implementations of verifier software (e.g. [17]) only provide a boolean decision regarding a system's integrity. A decision about the system's integrity is often not sufficient to ensure that a program's behaviour will be as expected [8]. For example, acceptable program behaviour may depend on specific compositions of system components and their versions.

Considering the above requirements, an authorisation policy for accessing sensitive data should be able to specify whether remote attestation is required or not. If required, it should also specify who the trusted verifier is and which conditions should be satisfied by system components in addition to the conditions over subject and data attributes.

Different organisations may want to specify different constraints over system components. Therefore, a boolean result from a common verification authority is not sufficient. Instead, a verifier should provide functionality to allow a data provider to identify attributes of components on the data consumer's machine in addition to verifying their integrity. These attributes can then be utilised to evaluate authorisation policies within an organisation as needed.

In this paper, we present a policy-driven approach that allows us to specify the remote attestation requirements as part of authorisation policies. We also describe our architecture (and its implementation) that is used to prove the integrity of the system. A Trusted Platform Service (TPS) was designed and developed to allow easy integration of secure applications with the remote attestation module. The decoupling of remote-attestation and authorisation logic allows for greater flexibility for an organisation to integrate data protection frameworks (such as Consequence [5]) with third-party attestation authorities. Our Trusted Platform service can be used by applications to handle remote attestation requests.

The major contribution of this paper is to illustrate the specification of remote attestation requirements as part of authorisation policies and to describe a modular implementation for easy integration of Trusted Computing technology with existing usage control systems. We believe that our approach and implementation on this platform will spur further investigation in this field.

The rest of the paper is organised as follows. Section 2 presents the overall architecture of our system. Section 3 presents the implementation details along with the various components that are used by our system. Section 4 presents research related to this paper, while Section 5 concludes the paper and provides ideas for future work.

2   Architecture

Evaluation of access rights on protected data requires identification of applicable policies and characteristics of the data. Therefore, metadata and policies are typically attached with the protected data during dissemination. In addition, the *content key* used for encrypting the data is encrypted with the public-key of a policy enforcement authority and also attached to the protected data. When a user requests access to the protected data, user and contextual attributes are also needed to evaluate the access request. These attribute values (as part of a credential / security token) must be provided by an authority trusted by the data provider or data owner.

Requirements for remote attestation of a system's integrity and state can be considered as contextual attributes and expressed as part of the usage control policies. An example policy using trusted credentials is shown below. Users can obtain the privileges for the "commander" role only if they present the required credentials. Access conditions can be further specified as part of authorisation policies (e.g. users must have an acceptable version of the glibc library).

```
authority A = "MyOrgCA"; authority V = "verifierCA";

credtype authn(uid, group);
credtype attestation(integrityCheck);
credtype version( glibcMajor, glibcMinor );

credential authnCred = authn signedby A;
credential attCred = attestation signedby V;
credential verCred = version signedby V;

role commander requires attCred.integrityCheck =="true"
and verCred and authnCred.group="officer" ;

authorization confidentialityOauthO = allow read()
    target( dataCategory == "personalSensitive" )
    to commander
    when ( (verCred.glibcMajor == 2 and verCred.glibcMinor>5)
        or verCred.glibcMajor>2 );
```

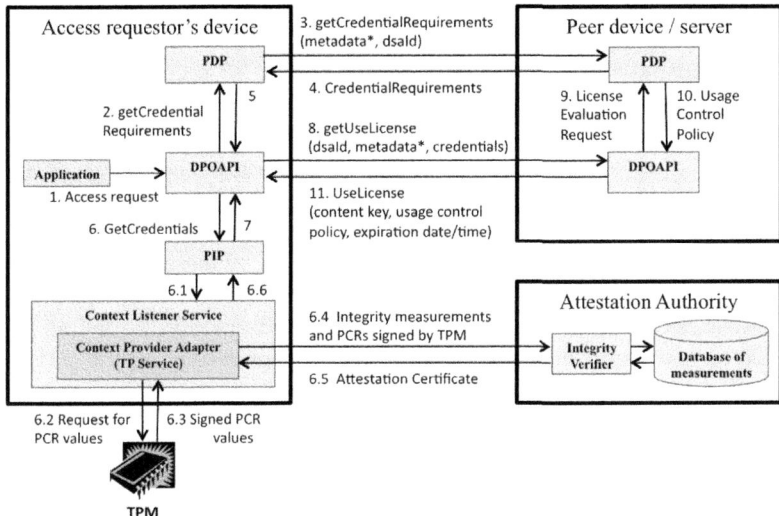

Fig. 1. Overall Architecture of the Remote Attestation Service

In Fig. 1, we show the interactions between components of our data protection framework. In this section, we describe how the framework can be integrated with a Trusted Platform Service (TPS). However, similar ideas may also be applied for integration with other dissemination control systems. The framework can be used to enforce *usage control* over data when data is shared within an organisation, and also if the data is shared across administrative domains using a *Data Sharing Agreement (DSA)*.

The Policy Enforcement Point (PEP) comprises a Data Protection Object (DPO) API and the application used to access the data. The DPO API is a generic enforcement component that can be used in different applications to enforce usage control. The framework contains both local and remote policy evaluation and enforcement components. The remote components play an important role in verifying credentials and evaluating policies for granting initial access to data. The local components are used to reevaluate policies and enforce a continuous control over the usage of data.

When a request to access protected data is made, the DPO API asks the Policy Decision Point (PDP) to determine the credentials needed to evaluate the access request. This is an optimisation step, as otherwise time may be spent on obtaining unnecessary credentials. The PDP then searches the policies associated with the protected data to determine the applicable ones on the basis of the metadata and requested action and determines the corresponding credential requirements. The credential requirements are expressed as pairs (credential type, issuer), where an issuer is the authority trusted to specify values for the credential type. For example, (authn, "MyOrgCA") is a requirement for a credential of the type *authn* that is signed by authority "MyOrgCA". The credential type is defined by a list of attribute names that must be present in a credential of that type. The issuer or authority value, e.g. "MyOrgCA" is a unique name for the authority used by the Policy Information Point (PIP) to identify it.

The PIP dispatches requests for the credentials to security token services. The architecture allows configuration of multiple *context adapters* into the PIP. Each context adapter can interact with a particular type of credential provider and obtain the requested credentials. The integration of the remote attestation process with the data protection framework has been realised by creating a specific context adapter.

If remote attestation is required, the DPO API requests the Trusted Platform Service (TPS) to obtain credentials from the specified verifier. The TPS collects the checksums values of all applications and modules loaded on the data consumer's machine (using the Integrity Measurement Architecture (IMA) [17]), along with a cumulative checksum maintained by the trusted hardware component called the Trusted Platform Module (TPM). The checksums are sent to the remote attestation authority specified in the given credential requirement. The verifier validates the checksums from the data consumer's machine against a database of known checksums for the trusted applications and system software. The attestation authority verifies that the measurements are from acceptable applications and then it calculates an expected value for the aggregate checksum based on the measurements presented. If the calculated aggregate checksum matches with the value signed by the data consumer's TPM, then the attestation authority knows that the data consumer's system has loaded only trusted components. In such a case, the attestation authority can issue a certificate for the successful integrity check of the data consumer's system. In addition, the verifier can attest specific information (such as version number) about components that were relevant for the policy evaluation.

In the previous example, the policy requires a minimum version (but not the latest version) for the glibc library. Such a policy may be useful as the applications accessing the data may have a vulnerability when using an older version of the library. The intention of such a policy is to capture organisation-specific security requirements that need not be enforced by the verifier. This separation of responsibilities between the verifier and the organisation specific authorisation policies is referred to as *decoupling of remote attestation logic from authorisation logic*.

After obtaining the required credentials, the DPO API requests the decryption key from the policy enforcement service, which verifies the credentials and asks a policy-service to check whether the access is authorised. As shown in Fig. 1, a successful evaluation leads to the issue of a use-license (which includes the decryption key).

Our approach has the advantage of mitigating the possibility of granting access to systems with known vulnerabilities in a way that does not require the attestation authority to know the organisation policy. After the use-license has been released, the enforcement module at the data consumer's machine protects decryption keys using the secure storage feature of the TPM. To ensure the correct functioning of the system, it is critical to ensure the integrity of the PDP,

DPO API and PIP components. These must be verified during a secure boot procedure as mentioned in Schmidt et al. [20].

To cope with a system's state change after attestation, we require re-attestation of the system to obtain a use-license for each data item and to renew any existing use-licenses, which expire at a time governed by the usage control policy. In our work, we do not specify an integrity metric, since one may need to analyse a huge number of software components and identify which ones are critical to system security. This only furthers our case for outsourcing the attestation task to a specialist.

3   Implementation

In this section, we present the background and implementation details of the architecture presented in Section 2.

3.1   Trusted Platform Module

The Trusted Platform Module is a micro-controller chip located on the motherboard and contains cryptographic engines and memory (both persistent and volatile). The components of a TPM are shown in Fig. 2. The TPM specification is an industry specification released by the Trusted Computing Group [23]. A TPM provides sealed storage and remote attestation capabilities. It performs cryptographic computations internally, i.e. hardware and software components outside the TPM do not have access to the execution of crypto functions within the TPM hardware.

Cryptographic operations are performed in a TPM by using a cryptographic accelerator, an engine for SHA-1, a HMAC engine, a Key Generator and a Random Number Generator. This allows RSA encryption and decryption and can also be used to sign data. The Platform Configuration Registers (PCRs) are each 20-bytes long and are used to store measurements (SHA-1 hash values) of the hardware and software configurations of the platform. A PCR r (with value r_t at time t) is updated with a new measurement m by padding m to the existing value in r and then taking the hash (using SHA-1) of the resultant value. In particular, $r_{t+1} = SHA - 1(r_t \parallel m)$.

To use the TPM functionalities such as TPM Quote (which produces a signed composite hash of the selected PCRs and external data, such as a nonce), the TPM needs to be set up with an Attestation Identity Key (AIK) and it's appropriate credentials. This key will be used to sign PCR values (such as from TPM Quote) and the associated credential will guarantee that the quote is coming from a genuine TPM. Using the platform credentials of the TPM, an AIK was created apriori by using the services provided by PrivacyCA [15].

3.2   Integrity Measurement Architecture

The runtime system of the device used by the requester must be attested to ensure its integrity. Attestation requires measurement of all components from the

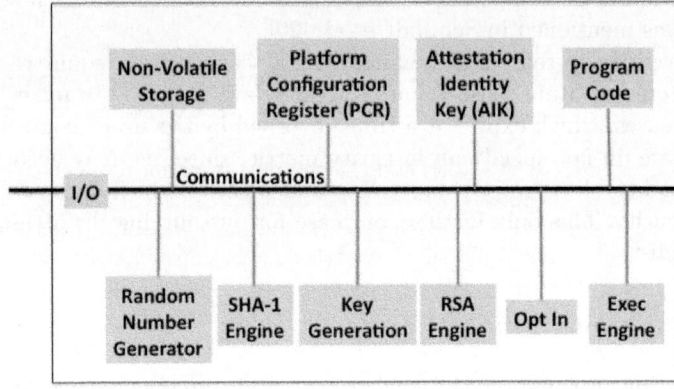

Fig. 2. Trusted Platform Module (TPM)

boot process up to the application layer. This is achieved by using the Integrity Measurement Architecture (IMA) [17]. To aid in taking measurements, hooks are placed in the Linux kernel. All executable content, as well as application-related file content (configuration files, libraries, etc.) are measured (a SHA-1 checksum of the file is taken) before they are loaded. The files measured include: kernel modules, executables, configuration input files. Each time a file is measured, a given PCR (*PCR-10*) is extended with the new value, and this value is also added to the measurement list that is stored in the kernel. This measurement list is also accessible using the filesystem. To attest a platform, the value of *PCR-10* along with its measurement list is sent to the verifier (Attestation Authority) who can check the state of the software stack. We use the Integrity Measurement Architecture that is available as part of the Linux Kernel (version 2.6.32).

3.3   Trusted Boot

A basic principle followed in trusted platform technologies is to verify the integrity or trust of every critical component before it is executed or loaded. Therefore, in addition to checking the runtime integrity of a system, we must also ensure that the system was booted correctly and is running the appropriate operating system. This is achieved by building a chain of trust starting with the Core Root of Trust for Measurement (CRTM), which is a trusted code in the BIOS boot block. It reliably measures integrity values of other entities, and stays unchanged during the lifetime of the platform. CRTM is an extension of normal BIOS which is run first to measure other parts of the BIOS block before passing control. The BIOS then measures hardware and the bootloader and passes control to the bootloader. The bootloader measures the OS kernel image and passes control to the OS. Each step of the boot process extends the appropriate PCR value in the TPM with the measurements taken in that step. These measurements attest the integrity of the system. For the purpose of our implementation, we used Trusted Grub v1.1.5 [24] as our boot loader and Linux kernel 2.6.32. This process is shown in Fig. 3.

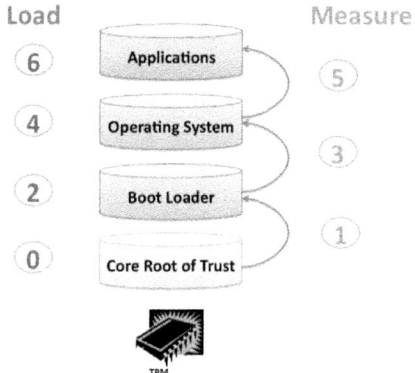

Fig. 3. Trusted Boot Process

3.4 Trusted Platform Service

The Trusted Platform Service (TPS) provides the enforcement architecture with
two functions to: (i) verify the integrity of the system and (ii) protect encryption
keys. The Trusted Platform Service is implemented in C and modelled as a
"listener" service that waits for requests on a particular port. It provides the
functionality of interacting with system-level libraries that measure system state
and is also responsible for interactions with the verifier. This service uses the
TrouSerS (Trusted Computing Software Stack) API [22] for accessing the various
functionalities of the TPM. The verification of the software stack is done using
the Trusted Platform Module, along with kernel and boot-loader improvements,
and the Integrity Measurement Architecture (IMA) [17].

When the Trusted Platform Service receives a request from the PIP, it initiates
the verification process. The request from the PIP contains information about the
verifier (including the name and port number) and using this the TPS contacts
the verifier. Upon receipt of a nonce (to ensure freshness) from the verifier, the
TPS packages the required data and sends it to the verifier. The packaged data
consists of: (i) TPM Quote (signed by the TPM), (ii) TPM credentials, (iii)
values of selected TPM PCRs (chosen depending on what we wish to check -
this includes *PCR-10*), and (iv) IMA measurement list.

3.5 Attestation Authority

The Attestation Authority is implemented in C and modelled as a "listener"
service that waits for requests on a particular port. The Attestation Authority
is provided with a database of expected checksum values of various programs. It
is against this list that incoming requests are checked. Upon receiving a request
from the Trusted Platform Service, the Attestation Authority replies back with
a nonce. The resultant response from the Trusted Platform Service is the data
blob containing the results of the TPM Quote function along with the TPM
credentials, as well as the IMA measurement list. The Attestation Authority

initially verifies the TPM Quote (using the credentials provided for the TPM), after which the PCR values are checked. To verify the integrity of the software stack, the IMA measurement is used to re-compute the value of *PCR-10* and it is compared to the received value. Finally, the hash values of the applications (and their versions) are checked against the available database.

Depending on the outcome of the verification process, the Attestation Authority will send back an Attestation certificate (signed by it to ensure authenticity) certifying the required credentials (as specified by the authorisation policy). We use X.509 certificates for this purpose. This method of attestation allows for decoupling between the evaluation of authorisation policies (at the data consumer's side) and remote-attestation at the Attestation Authority.

4   Related Work

Remote attestation is an integral part of Trusted Computing and the ability to verify the software and/or hardware running on a machine is of paramount importance. There has been however, some criticism about whether this is a viable or impractical solution, since it requires the attesting authority (or verifier) to know apriori the needed software configurations as well as their checksums. Lyle et al. show that this is indeed a viable solution for web services [11]. Our remote attestation scheme is generic enough to work with web services as well as other technologies.

The Integrity measurement architecture (IMA) proposed by Sailer et al. [17] uses binary attestation to measure all the programs and code before they are loaded into the system to run. This architecture was extended by Jaeger et al. in [9], where the proposed architecture only measures the programs and code that are needed by the verifier. This was due to the fact that the verifier does not necessarily need to know the measurements of all the programs that are running on the system. In our work, we use IMA because we do not distinguish between what is required by the verifier and that which is required by the enforcement layer. This is due to the fact that these two are decoupled and different third-party verifiers could be used for our purpose.

In the event that the software platform cannot guarantee that the software it is running is reliable, it is advisable to move the required program code away from the untrusted platform. The technique used for this is called code-splicing, which involves splitting the program code into "critical" and "non-critical" sections, so that only non-critical code is run on the untrusted platform. This in turn guarantees that code that is critical has not been tampered with. These ideas have been presented by Ceccato et al. [3], Dvir et al. [6] and Zhang et al. [27].

Kennell et al. proposed a system called "Genuinity", which verifies if the hardware and the software running on a system are "genuine" [10]. This is achieved without the use of any special hardware and by using a timed execution of a checksum function that provides a fingerprint of the running applications. The time taken to execute the checksum is verified by the verifier. As shown by Shanker et al. [21], this is not a viable solution due to the assumptions that are to be imposed and it is also prone to substitution attacks.

Using the idea of calculating a checksum, Schellekens et al. [19] proposed a system where the time stamping functionality of the TPM is used to calculate the execution time of the checksum locally. This is then used in conjunction with a timing based remote attestation mechanism to prove the integrity of the system.

Alawneh et al. propose an architecture to protect data within an organisation [1]. Their threat model is that of a rogue employee who can potentially disseminate data to the outside world. The proposed system binds the sensitive content within the organisation to specific devices, thereby restricting the content from being leaked to other devices. In the case when devices need to share information, these devices are allocated to the same dynamic domain.

Although the aforementioned systems all provide varying degrees of remote attestation and verification, none of them decouples the attestation process from the enforcement process. This allows for more flexibility with respect to the type of remote attestation chosen, as well as more fine grained access control at the enforcement layer.

Sadeghi et al. [16] provide reasons as to why binary attestation may not be the most useful form of attestation. They argue that rather than using the checksums of the software programs on a machine, it is better to asses the state of a platform based on some pre-determined security properties. This type of attestation that uses the security property rather than binary attestation is referred to as property-based attestation. Property-based attestation uses binary attestation to prove some "property" of the system (for example, if certain hash values match, then we can state that the system is in a particular state). There have been several techniques that have been proposed for property-based attestation [4,13]. In [12], the authors have proposed the idea of a Property Manifest, which can be used to define security policies for policy-based attestation.

Haldar et al. proposed the concept of semantic remote attestation in [8], wherein the security of the system is guaranteed through program analysis. To verify whether a program's execution will satisfy the security properties, the authors analyse the program by using a Trusted Virtual Machine.

Though both property-based attestation and semantic remote attestation provide the techniques to attest whether a remote machine is trusted or not, neither of these decouple the enforcement layer from the attestation authority. Also, this decoupling allows our system to use any underlying attestation mechanism and this adds to the flexibility of the presented system.

In [26], Yu et al. propose a system for guaranteeing the freshness of the integrity measurement that is used in the attestation. This proposed solution, called RTRA, uses the run-time state of the attesting party. RTRA can be directly integrated into our architecture by using it as the underlying attestation mechanism.

Often privacy is also a concern when sharing system measurements with the verifier. Our scheme may be used together with privacy preserving protocols such as Brickell et al.'s attestation scheme [2]. The high-level description of their scheme implies changes in the cryptographic protocol between IMA, TPM, and the verifier. However, our architecture resides at a higher-level and is independent of the cryptography specifics used (e.g. signature schemes).

The goals of Trusted Network Connect [25] seem very similar to our work. In their architecture the integrity measurement verifiers act as the PDP, which is different from our architecture as we separate the PDP and verifier. They also have the problem of policies being evaluated by the verifier itself.

5   Conclusion and Future Work

In this paper, we have presented a policy-driven approach that allows us to prove the integrity of a system while decoupling authorisation logic from remote attestation. This decoupling of remote attestation and authorisation logic allows for greater flexibility for an organisation to fine tune their authorisation policies by using different data protection frameworks with third-party attestation authorities. This system relies on a trusted hardware component, the Trusted Platform Module and uses the Integrity Measurement Architecture. In the future, we would like to evaluate our system with different policy-based data protection frameworks as well as with different attestation mechanisms.

References

1. Alawneh, M., Abbadi, I.M.: Sharing but protecting content against internal leakage for organisations. In: DBSec, pp. 238–253 (2008)
2. Brickell, E., Camenisch, J., Chen, L.: Direct anonymous attestation. In: Proceedings of the 11th ACM Conference on Computer and Communications Security, CCS 2004, pp. 132–145. ACM, New York (2004)
3. Ceccato, M., Preda, M., Nagra, J., Collberg, C., Tonella, P.: Barrier slicing for remote software trusting. In: Seventh IEEE International Working Conference on Source Code Analysis and Manipulation, SCAM 2007, September 30-October 1, pp. 27–36 (2007)
4. Chen, L., Landfermann, R., Löhr, H., Rohe, M., Sadeghi, A.-R., Stüble, C.: A protocol for property-based attestation. In: Proceedings of the First ACM Workshop on Scalable Trusted Computing, STC 2006, New York, NY, USA, pp. 7–16 (2006)
5. Consequence Project, http://www.consequence-project.eu/
6. Dvir, O., Herlihy, M., Shavit, N.: Virtual leashing: Internet-based software piracy protection. In: Proceedings of 25th IEEE International Conference on Distributed Computing Systems, ICDCS 2005, pp. 283–292 (June 2005)
7. Gowadia, V., Scalavino, E., Lupu, E.C., Starostin, D., Orlov, A.: Secure cross-domain data sharing architecture for crisis management. In: Proceedings of the Tenth Annual ACM Workshop on Digital Rights Management, DRM 2010, New York, NY, USA, pp. 43–46 (2010)
8. Haldar, V., Chandra, D., Franz, M.: Semantic remote attestation - a virtual machine directed approach to trusted computing. In: USENIX Virtual Machine Research and Technology Symposium, pp. 29–41 (2004)
9. Jaeger, T., Sailer, R., Shankar, U.: Prima: Policy-reduced integrity measurement architecture. In: Proceedings of the Eleventh ACM Symposium on Access Control Models and Technologies, SACMAT 2006, New York, NY, USA, pp. 19–28 (2006)

10. Kennell, R., Jamieson, L.H.: Establishing the genuinity of remote computer systems. In: Proceedings of the 12th Conference on USENIX Security Symposium, vol. 12, pages 21. USENIX Association, Berkeley (2003)
11. Lyle, J., Martin, A.: On the feasibility of remote attestation for web services. In: International Conference on Computational Science and Engineering, CSE 2009, vol. 3, pp. 283–288 (August 2009)
12. Nagarajan, A., Varadharajan, V., Hitchens, M., Arora, S.: On the applicability of trusted computing in distributed authorization using web services. In: DBSec, pp. 222–237 (2008)
13. Nagarajan, A., Varadharajan, V., Hitchens, M., Gallery, E.: Property based attestation and trusted computing: Analysis and challenges. In: NSS, pp. 278–285 (2009)
14. Park, J., Sandhu, R.S., Schifalacqua, J.: Security architectures for controlled digital information dissemination. In: Proc. of ACSAC, p. 224 (2000)
15. PrivacyCA, http://www.privacyca.com/
16. Sadeghi, A.-R., Stüble, C.: Property-based attestation for computing platforms: caring about properties, not mechanisms. In: Proceedings of the 2004 Workshop on New Security Paradigms, NSPW 2004, New York, NY, USA, pp. 67–77 (2004)
17. Sailer, R., Zhang, X., Jaeger, T., van Doorn, L.: Design and implementation of a tcg-based integrity measurement architecture. In: Proceedings of the 13th Conference on USENIX Security Symposium, SSYM 2004, vol. 13, pages 16. USENIX Association, Berkeley (2004)
18. Sandhu, R.S., Ranganathan, K., Zhang, X.: Secure information sharing enabled by Trusted Computing and PEI models. In: ASIA CCS, pp. 2–12 (2006)
19. Schellekens, D., Wyseur, B., Preneel, B.: Remote attestation on legacy operating systems with trusted platform modules. Sci. Comput. Program 74, 13–22 (2008)
20. Schmidt, A.U., Leicher, A., Cha, I., Shah, Y.: Trusted platform validation and management. International Journal of Dependable and Trustworthy Information Systems (IJDTIS) 1(2), 1–31 (2010)
21. Shankar, U., Chew, M., Tygar, J.D.: Side effects are not sufficient to authenticate software. In: Proceedings of the 13th USENIX Security Symposium, pp. 89–101 (2004)
22. TrouSerS - The open-source TCG Software Stack, http://trousers.sourceforge.net/
23. Trusted Computing Group, http://www.trustedcomputinggroup.org/
24. Trusted Grub, http://sourceforge.net/projects/trustedgrub/
25. Trusted Network Connect, http://www.trustedcomputinggroup.org/files/resource_files/51F9691E-1D09-3519-AD1C1E27D285F03B/TNC_Architecture_v1_4_r4.pdf
26. Yu, A., Feng, D.: Real-Time Remote Attestation with Privacy Protection. In: Katsikas, S., Lopez, J., Soriano, M. (eds.) TrustBus 2010. LNCS, vol. 6264, pp. 81–92. Springer, Heidelberg (2010)
27. Zhang, X., Gupta, R.: Hiding program slices for software security. In: Proceedings of the International Symposium on Code Generation and Optimization: Feedback-Directed and Runtime Optimization, CGO 2003, pp. 325–336. IEEE Computer Society, Washington, DC (2003)

ID-Based Deniable Authentication Protocol Suitable for Mobile Devices

Jayaprakash Kar

Department of Information Systems,
Faculty of Computing and Information Technology,
King Abdulaziz University, Kingdom of Saudi Arabia
jayaprakashkar@yahoo.com

Abstract. This paper describes a secure identity based deniable authentication protocol whose security is based on difficulty of breaking Diffie-Hellman Problem on Elliptic Curve (ECDHP) and hash function. Elliptic curve cryptosystem (ECC) has significant advantages like smaller key sizes, faster computations compared with other public-key cryptography. Since it is an ECC based authentication protocol, it can be implimented in mobile devices such as smart card, PDA etc. Deniable authentication protocol enables a receiver to identify the true source of a given message, but not to prove the identity of the sender to a third party. This property is very useful for providing secure negotiation over the Internet.

Keywords: deniable authentication, ECDLP, ECDHP, HDDH, mobile device.

1   Introduction

Authentication can be realized by the use of digital signature in which the signature (signers private key) is tied to the signer as well as the message being signed. This digital signature can later be verified easily by using the signers public key. Hence, the signer will not be able to deny his articipation in this communication. Generally, this notion is known as non-repudiation. However, under certain circumstances such as electronic voting system, online shopping and negotiation over the Internet, the non-repudiation property is undesirable. It is important to note that in these applications, the senders identity should be revealed only to the intended receiver. Therefore, a significant requirement for the protocol is to enable a receiver to identify the source of a given message, and at the same time, unable to convince to a third party on the identity of the sender even if the receiver reveal his own secret key to the third party. This protocol is known as deniable authentication protocol.

2   Applications to Mobile Devices

With the rapid development of the development of electronic technology, various mobile devices (e.g., cell phone, PDA, and notebook PC) are produced and

R. Prasad et al. (Eds.): MOBISEC 2011, LNICST 94, pp. 160–171, 2012.

peoples life is made more convenient. More and more electronic transactions for mobile devices are implemented on Internet or wireless networks. In electronic transactions, remote user authentication in insecure channel is an important issue. For example, when one user wants to login a remote server and access its services, such as on-line shopping, both the user and the server must authenticate the identity with each other for the fair transaction. Generally, the remote user authentication can be implemented by the traditional public-key cryptography (Rivest et al., 1978; ElGama, l985). The computation ability and battery capacity of mobile devices are limited, so traditional public-key cryptograph, in which the computation of modular exponentiation is needed, cant be used in mobile devices.

Fortunately, Elliptic Curve Cryptosystem (ECC) (Miller, 1986; Koblitz, 1987) has significant advantages like smaller key sizes, faster computations compared with other public-key cryptography. Thus, ECC-based authentication protocols are more suitable for mobile devices than other cryptosystem. However, like other public-key cryptography, ECC also needs a key authentication center (KAC) to maintain the certificates for users public keys. When the number of users is increased, KAC needs a large storage space to store users public keys and certificates.

ECC has the highest strength-per-bit compared to other public key cryptosystems. Small key sizes translate into savings in bandwidth, memory and processing power. This makes ECC the obvious choice in this situation. However, there are other aspects that need to be taken into account. When it comes to choosing which public key cryptosystem to employ in a mobile environment, one has to keep in mind restrictions on bandwidth, memory and battery life. In constrained environments such as mobile phones, wireless pagers or PDAs, these resources are highly limited. Thus, a suitable public key scheme would be one that is efficient in terms of computing costs and key sizes.

When it comes to choosing which public key cryptosystem to employ in a mobile environment, one has to keep in mind restrictions on bandwidth, memory and battery life. In constrained environments such as mobile phones,wireless pagers or PDAs, these resources are highly limited. Thus, a suitable public key scheme would be one that is efficient in terms of computing costs and key sizes. This protocol can be implemented in low power and small processor mobile devices such as smart card, PDA etc which work in low power and small processor. Since the proposed protocol is based on ECC, can be implimented to mobile devices especially Smart card.

3   Preliminaries

3.1   Notations

We first introduce common notations used in this paper as follows.

- p is the order of underlying finite field;
- F_p is the underlying finite field of order p

- E is an an elliptic curve defined on finite field F_p with large order.
- G is the group of elliptic curve points on E.
- P is a point in $E(F_p)$ with order n , where n is a large prime number.
- $\mathcal{H}(\cdot)$ is a secure one-way hash function.
- $\|$ denotes concatenation operation between two bit stings.
- S be the Sender with identity ID_s, $ID_s \in \{0,1\}^*$.
- R be the Receiver with identity ID_r, $ID_r \in \{0,1\}^*$.

4   Diffie-Hellman Problem

This section briefs overview of Computational Diffie-Hellman (CDH) problem, Decisional Diffie-Hellman and Hash Diffie-Hellman problem in $\mathbb{G}$.

Definition 1. Diffie-Hellman Problem: *Let $(q, \mathbb{G}, P)$ be a 3-tuple generated by polynomial time algorithm $\mathcal{G}(k)$, and let $a, b \in \mathbb{Z}_q^*$, the CDH problem in $\mathbb{G}$ is as follows: Given (P, aP, bP), compute abP. The (t, ϵ)-CDH assumption holds in $\mathbb{G}$ if there is no algorithm $\mathcal{A}$ running in time t such that*

$$\mathbf{Adv}_{\mathbb{G}}^{CDH}(\mathcal{A}) = Pr[\mathcal{A}(P, aP, bP) = abP] \geq \epsilon$$

where the probability is taken over all possible choices of (a, b).

$\underline{\mathbf{Exp}_{\mathcal{G}(k)}^{CDH}}$

1. $(\mathbb{G}, q, P) \leftarrow \mathcal{G}(1^k)$
2. $a, b, c \leftarrow \mathbb{Z}_q^*$
3. $U_1 = aP, U_2 = bP$
4. if $W = abP$ return 1 else return 0

Definition 2. Decisional Diffie-Hellman Problem: *Let $(q, \mathbb{G}, P)$ be a 3-tuple generated by polynomial time algorithm $\mathcal{G}(k)$, and let $a, b, c \in \mathbb{Z}_q^*$, the DDH problem in $\mathbb{G}$ is as follows: Given (P, aP, bP, cP), decide whether it is a Diffie-Hellman tuple.*

Definition 3. Hash Decisional Diffie-Hellman Problem: *Let $(q, \mathbb{G}, P)$ be a 3-tuple generated by polynomial time algorithm $\mathcal{G}(k)$, $\mathcal{H} : \{0,1\}^* \rightarrow \{0,1\}^l$ be a secure cryptographic hash function, whether l is a security parameter, and let $a, b \in \mathbb{Z}_q^*, h \in \{0,1\}^l$, the HDDH problem in $\mathbb{G}$ is as follows: Given (P, aP, bP, h), decide whether it is a hash Diffie-Hellman tuple $((P, aP, bP, \mathcal{H}(abP))$. If it is right, outputs 1; and 0 otherwise. The (t, ϵ)-HDDH assumption holds in $\mathcal{G}$ if there is no algorithm $\mathcal{A}$ running in time at most t such that*

$$\mathbf{Adv}_{\mathbb{G}}^{HDDH}(\mathcal{A}) = |Pr[\mathcal{A}(P, aP, bP, \mathcal{H}(abP) = 1] - Pr[\mathcal{A}(P, aP, bP, h) = 1]| \geq \epsilon$$

where the probability is taken over all possible choices of (a, b, h).

5   Deniable Property

Deniable authentication protocol is a new security authentication mechanism. Compared with traditional authentication protocols, it has the following two features:

1. It enables an intended receiver to identify the source of a given message.
2. However, the intended receiver can not prove to any third party the identity of the sender

In 1998, Dwork et al. [10] developed a notable deniable authentication protocol based on the concurrent zero-knowledge proof, however the protocol requires a timing constraint and the proof zero-knowledge is subject to a time delay in the authentication process. Auman and Rabin [11] proposed some other deniable authentication protocols based on the factoring problem. In 2001, Deng et al. [15] also proposed two deniable authentication protocols based on the factoring and the discrete logarithm problem respectively.

The proposed protocol will be achieving the following properties.

- **Deniable Authentication:** The intended receiver can identify the source of a given message, but cannot prove the source to any third party.
- **Authentication:** During the protocol execution, the sender and the intended receiver can authentication each other.
- **Confidentiality:** Any outside adversary has no ability to gain the deniable authentication message from the transmitted transcripts.

6   Security Model

Security Notions In this subsection, we explain the security notions ofID-based deniable authentication protocol. We first recall the usual security notion: the unforgeability against chosen message attacks (Goldwasser et al., 1988), then we consider another security notion: the deniablity of deniable authentication protocol [2].

Player. Let $P = \{\mathcal{P}_0, \mathcal{P}_1, \ldots \mathcal{P}_n\}$ be a set of players who may be included in the system. Each player $\mathcal{P}_i \in P$ get his public-secret key pair (pk_i, sk_i) by providing his identity i to the **Extract** algorithm. A player $\mathcal{P}_i \in P$ is said to be fresh if $\mathcal{P}_i$'s secret key sk_i has not been revealed by an adversary; while if $\mathcal{P}_i$s secret key sk_i has been revealed, $\mathcal{P}_i$ is then said to be corrupted. With regard of the unforgeability against chosen-message attacks, we define the security notion via the following game played by a challenger and an adversary.

[**Game 1**]

- Initial: The challenger runs Setup to produce a pair $(params, master - key)$, gives the resulting $params$ to the adversary and keeps the master-key secretly.

- Probing: The challenger is probed by the adversary who makes the following queries.
- Extract: The challenger first sets $\mathcal{P}_0, \mathcal{P}_1$ to be fresh players, which means that the adversary is not allowed to make Extract query on $\mathcal{P}_0$ or $\mathcal{P}_1$. Then, when the adversary submits an identity i of player $\mathcal{P}_i, (i = 0, 1)$, to the challenger. The challenger responds with the public-secret key pair (pk_i, sk_i) corresponding to i to the adversary.
- Send: The adversary submits the requests of deniable authentication messages between $\mathcal{P}_0$ and $\mathcal{P}_0$. The challenger responds with deniable authentication messages with respect to $\mathcal{P}_0$ (resp. $\mathcal{P}_1$) to $\mathcal{P}_1$ (resp $\mathcal{P}_0$).
- Forging: Eventually, the adversary outputs a valid forgery $\tilde{m}$ between $\mathcal{P}_0$ and $\mathcal{P}_1$. If the valid forgery $\tilde{m}$ was not the output of a Send query made during the game, we say the adversary wins the game.

Definition 4. (Unforgeability). *Let A denote an adversary that plays the game above. If the quantity $Adv^{UF}_{IBDAP}[A] = Pr[Awins]$ is negligible we say that the ID-based deniable authentication protocol in question is existentially unforgeable against adaptive chosen-message attacks.*

To capture the property of deniablity of deniable authentication protocol, we consider the following game run by a challenger.

[Game 2]

- Initial: Let $\mathcal{P}_0$ and $\mathcal{P}_1$ be two honest players that follow the deniable authentication protocol, and let $\mathcal{D}$ be the distinguisher that is involved in the game with $\mathcal{P}_0$ and $\mathcal{P}_0$.
- Challenging: The distinguisher $\mathcal{D}$ submits a message $m \in \{0,1\}^*$ to the challenger. The challenger first randomly chooses a bit $b' \in \{0,1\}^*$, then invokes the player P_b to make a deniable authentication message $\tilde{m}$ on m between $\mathcal{P}_0$ and $\mathcal{P}_1$. In the end, the challenger returns $\tilde{m}$ to the distinguisher $\mathcal{D}$.
- Guessing: The distinguisher $\mathcal{D}$ returns a bit $b \in \{0,1\}^*$. We say that the distinguisher $\mathcal{D}$ wins the game if $b = b'$.

Definition 5. (Deniablity). *Let D denote the distinguisher that is involved the game above. If the quantity $Adv^{DN}_{IBDAP}[D] = |Pr[b = b'] - \frac{1}{2}|$ is negligible we say that the ID-based deniable authentication protocol in question is deniable.*

7   Proposed Protocol

The Protocol follows the followings steps.

- **Setup.** Let $\mathcal{H} : \{0,1\}^* \to \{0,1\}^l$ be a secure cryptographic hash function which is of collision free. In the proposed protocol the sender has a certificate issued by the certificate authority (CA). The CA contains the public key (π_{pub}) of the sender, and the signature of CA for the certificate. The receiver can obtain (π_{pub}) and verify the validity of it. The private key (π_{prv}) of sender is kept secret.

− **Extract.** During the extraction phase, the sender S with identity $ID_s \in \{0,1\}^*$ select t_s randomly from $[1, n-1]$ and computes the following

$$a_s = \mathcal{H}(ID_s) \oplus t_s \tag{1}$$

$$Q_s = a_s \cdot P \tag{2}$$

The key pair is (Q_s, a_s). Then concatenate Q_s with the time stamp $T \in \mathbb{Z}_q^*$. Encrypts the concatenated value $(Q_s \| T)$ using his own private key π_{prv}.

$$\tilde{Q}_s = E_{\pi_{prv}}(Q_s \| T)$$

Similarly the receiver R with identity $ID_r \in \{0,1\}^*$ selects random number $t_r \in [1, n-1]$. Then computes the following:

$$a_r = \mathcal{H}(ID_r) \oplus t_r \tag{3}$$

$$Q_r = a_r \cdot P \tag{4}$$

So the key pairs of receiver R is (a_r, Q_r).
− **Send.** It follows the following steps.
   1. **Step 1:** During this phase the sender S sends the cipher $\tilde{Q}_s$ to the the receiver R. After getting, R will decrypt using the public key π_{pub} as $Q_s = D_{\pi_{pub}}(\tilde{Q}_s)$, where D denotes decryption algorithm.
   2. **Step 2:** Receiver R use the calculated value a_r from Eq.(3) and computes the session key α_1 as by the following equation.

$$\alpha_1 = a_r \cdot Q_s \tag{5}$$

   Receiver R sends the computed Q_r to S. Similarly Sender also compute the session key as

$$\alpha_2 = a_s \cdot Q_r \tag{6}$$

   In fact $\alpha = \alpha_1 = a_r \cdot Q_s = a_r a_s \cdot P = a_s \cdot Q_r = \alpha_2$
   3. **Step 3:** When Sender S authenticates the deniable message $M \in \{0,1\}^l$, computes $\gamma_1 = \mathcal{H}(\alpha_2, M \| T)$.
   4. **Step 4:** The resulting deniable authenticated message is tuples $\psi = (ID_s, T, \gamma_1)$.
   5. **Step 5:** Finally S sends ψ to the recipient R.
− **Receive**
   1. **Step 1:** After receiving $\psi = (ID_s, T, \gamma_1)$, the recipient R computes $\gamma_2 = \mathcal{H}(\alpha_1, M \| T)$
   2. **Step 2:** If the time stamp T is valid and $\gamma_1 = \gamma_2$, accepts M otherwise reject.

The protocol is illustrated in the following fig.

Sender S	Receiver R

Select random number
$t_s \in [1, n-1]$
Computes $a_s = \mathcal{H}(ID_s) \oplus t_s$
where $ID_s \in \{0,1\}^*$
Computes $Q_s = a_s \cdot P$
Encrypt Q_s as
$\tilde{Q}_s = E_{\pi_{prv}}(Q_s \| T)$
where $T \in \mathbb{Z}^*$
is the time stamp

$$\tilde{Q}_s \longrightarrow$$

Decrypt as $D_{\pi_{pub}}(\tilde{Q}_s \| T) = Q_s$
Select random number
$t_r \in [1, n-1]$
$a_r = \mathcal{H}(ID_r) \oplus t_r$
where $ID_r \in \{0,1\}^*$
Computes $\alpha_1 = a_r \cdot Q_s$
Computes $Q_r = a_r \cdot P$

$$Q_r \longleftarrow$$

Compute $\alpha_2 = a_s \cdot Q_r$
and $\gamma_1 = \mathcal{H}(\alpha_2, M \| T)$
$\psi = (ID_s, T, \gamma_1)$

$$\psi \longrightarrow$$

Computes $\gamma_2 = \mathcal{H}(\alpha_1, M \| T)$
if time stamp T is valid
and $\gamma_1 = \gamma_2$
accept M
otherwise reject

8   Correctness

Theorem 1. *If $\psi = (ID_s, T, \gamma_1)$ is a authentication message produced by the Sender S honestly, then the recipient R will always accept it.*

Proof: The proposed protocol satisfies the property of correctness. In effect, if the deniable authetication message ψ is correctly generated, then

$$\gamma_1 = \mathcal{H}(\alpha_2, M \| T) = \mathcal{H}(\alpha_1, M \| T) = \gamma_2$$
$$\text{Since } \alpha_1 = a_r \cdot Q_s = a_r a_s \cdot P = a_s \cdot Q_r = \alpha_2$$

9   Security Analysis

In this section, we analyze the security of our proposed deniable authentication protocol. The security of our protocol is based on Computational Diffie-Hellman (CDH), Decisional Diffie-Hellman (DDH) and the Hashed Diffie-Hellman (HDDH) Problems.

9.1   Security Model

The protocol is defined by the following game between an adversary A and a challenge C

- **Setup:** On input of security parameters, C runs the algorithm to generate the system parameters and public key and private key pairs $(pk_i, sk_i), 1 \leq i \leq n$, of n users $\{U = U_1, U_2, \ldots U_n\}$, and sends the system parameters and all public keys $pk_1, pk_2 \ldots pk_n$ to A.
- **Corrupt Queries:** A can corrupt some users in U and obtain their private keys.
- **User Authentication Queries:** A also can make several user authentication queries on some uncorrupted users in U.
- **Impersonate:** In the end, A impersonates an uncorrupted user in U by outputting a valid login authentication message.

The success probability of A to win the game is defined by **Succ(A)**.

Definition 6. *A user authentication scheme is secure if the probability of success of any polynomial bounded adversary A in the above game is negligible.*

Theorem 2. *Assume that the collision-free hash function $\mathcal{H}$ behaves as a random oracle. Then the proposed authentication scheme is secure provided that the Diffie-Hellman algorithm assumption holds in $\mathbb{G}$.*

Proof: Assume that A is an adversary, who can with non-negligible probability, break the proposed authentication scheme. Then, we can use A to construct another algorithm $\tilde{A}$, which is having parameters $(q, \mathbb{G}, P)$ and $\mathcal{H}$, where $\mathcal{H} : \{0,1\}^* \rightarrow \{0,1\}^l$ be a secure cryptographic hash function, behaves a random oracle [7] and a DH instance (P, aP, bP), where $a, b \in \mathbb{Z}_q^*$ as her challenge, and her task here is to compute $(ab) \cdot P$. Let $U = U_1, U_2 \ldots U_n$ be a set of n users who may participate in the system. $\tilde{A}$ first picks a random number j from $\{1, 2 \ldots n\}$, and sets the user U_j's public key $Q_j = t_j \cdot P$. Then, $\tilde{A}$ chooses another $n - 1$ random numbers $t_i \in \mathbb{Z}_q^*$ as user U_i's secret key, where $1 \leq i \leq n$ and $i \neq j$, and computes the corresponding public key $Q_i = t_i \cdot P$. Finally, $\tilde{A}$ sends all public key $Q_1, Q_2 \ldots Q_n$ to the adversary A.

Theorem 3. *The proposed Protocol achieves the authentication between the sender and the intended receiver.*

Proof: In our proposed protocol, if the receiver accepts the authentication message ψ, receiver R can always identify the source of the message. If an adversary wants impersonate the sender S, he can obtain a time stamp $T \in \mathbb{Z}_q^*$, a message M. But, he could not construct the α_2. If the adversary tries to compute α_2 he has to know the sender's private key a_s for that it needs to solve ECDLP.

Definition 7. *Informally, a deniable authentication protocol is said to achieve the property of confidentiality, if there is no polynomial time algorithm that can distinguish the transcripts of two distinct messages.*

Theorem 4. *The proposed protocol achieves the property of confidentiality provided that the HDDH problem is hard in* $\mathbb{G}$.

Proof : $\gamma_1 = \mathcal{H}(\alpha_2, M \| T)$ is actually a hashed ElGamal cipher text [14]. Hashed ElGamal encryption is semantically secure in the random oracle model under the Computational Diffie-Hellman (CDH) assumption. This is the assumption that given P, aP, bP, it is hard to compute $ab \cdot P$ in $\mathbb{G}$, where a, b are random elements of $\mathbb{Z}_q^*$. The CDH assumption is more precisely formulated as follows.
Let $\mathcal{A}$ be an algorithm that takes as input a pair of group elements, and outputs a group element. CDH-advantage of $\mathcal{A}$ to be

$$[a, b \leftarrow \mathbb{Z}_q^* : \mathcal{A}(aP, bP) = ab \cdot P]$$

The CDH assumption on $(\mathbb{G})$ is that any efficient algorithms CDH advantage is negligible. As a result, the proposed protocol can achieves the confidentiality.

Theorem 5. *The proposed protocol also achieves the property of deniability.*

Proof : To prove that the proposed protocol has deniable property, first we should prove that it enables an intended receiver R to identify the source of the given message M. Since the authenticated message $\psi = (ID_s, T, \gamma_1)$ contains the sender identity ID_s, R can easily identify the source of the message. After verifying $\gamma_1 = \gamma_2$, R can be assured that the message is originated from S. If R intends to expose the message's identity to third party, S would be repudiate as he would argue that S could also generate ψ, since R can compute γ_2 and $\gamma_1 = \gamma_2$, *i.e* transcripts transmitted between the sender S and the receiver R could be simulated by the receiver R himself in polynomial time algorithm. Hence the deniable property is satisfied.
    Also we can prove considering the security model describe in section-5. Let us consider a distinguisher $\mathcal{D}$ and two honest players $\mathcal{P}_0$ and $\mathcal{P}_1$ involved in **Game** 2. The distinguisher $\mathcal{D}$ first submits a message $m \in \{0,1\}^*$ to the challenger. Then, the challenger chooses a bit $b \in \{0,1\}$ uniformly at random, and invokes the player $\mathcal{P}_b$ to make a deniable authentication message $\psi = (ID_b, T_b, MAC_b, C)$ on m between $\mathcal{P}_0$ and $\mathcal{P}_1$. In the end, the challenger returns $\psi = (ID_b, T_b, MAC_b, C)$ to the distinguisher $\mathcal{D}$. Since both $\mathcal{P}_0$ and $\mathcal{P}_1$ can generate a valid deniable authentication message $\psi = (ID_b, T_b, MAC_b, C)$, which can pass the verification equation, in an indistinguishable way, when $\mathcal{D}$ returns the guessed value b, we can sure that the probability $\Pr[b = b']$ is $\frac{1}{2}$, and the quantity $Adv_{IBDAP}^{DN}[D] = |Pr[b = b'] - \frac{1}{2}| = |\frac{1}{2} - \frac{1}{2}| = 0$ Based upon

the analysis above, we can conclude that the proposed protocol can achieve the deniable authentication.

Theorem 6. *The Protocol authenticates the source of the message.*

Proof: If someone proves $\mathcal{H}(\alpha_2, M\|T)$ to R, where $\alpha_2 = a_s \cdot Q_r$, he must be S. If an adversary gets all the information Q_s in **Extract** phase, he can not compute the session key α_1. It is as difficult as solving Elliptic Curve Discrete Logarithm Problem.

Definition 8. Secure against Man-in-the-middle. *An authentication protocol is secure against an Man-in-the-middle, if Man-in-the-middle can not establish any session key with either the sender or the receiver.*

Theorem 7. *The proposed protocol is secure with respect to the man-in-the-middle (MIA) attack.*

Proof: In the extraction phase, the message is encrypted with the private key π_{prv}. It is difficult for the adversary to get the key π_{prv}. An intruder can intercept the message from S and act as R to negotiate the session key α with S. If he wants execute MIA attack, he must act as the sender S to cheat R. To construct the cipher $\tilde{Q}_s$, first he has to find out π_{prv} and a_s. For that he has to solve ECDLP, which is computationally infeasible takes fully exponential time. If he fakes an $\tilde{Q}_s$, R can not get correct Q_s. so it resist MIA attack.

10   Computational Complexity

The computation cost for the performance of this new protocol is as follows: the sender needs to compute a point multiplication, a pairing evaluation, an encryption, as well as a hash evaluation. In addition, the most expensive work for the sender is the use of a public-key digital signature algorithm.Since the receiver and the sender stand in the symmetric position, so the receiver shares the same computation costs. The communication cost of the proposed protocol is that the sender and the receiver carry out two rounds for communications in order for the receiver to obtain a message from the sender.

Let T_M : is the time taken for executing a scalar multipication over Elliptic Curve.
T_H : is the time for executing one-way hash function.
$T_\oplus$: is the time taken for Exclusive OR operation.
$T_{Encyp\&Decrp}$: is the time taken for encryption and decryption in Public Key cryptosystem.
Execution time, Sender has to take is $T_S = 2T_M + 2T_H + T_\oplus + T_{Encrp}$.
Execution time, Receiver has to take is $T_R = 2T_M + 2T_H + T_\oplus + T_{Decrp}$.
Total time $T = T_S + T_R$.

In practical implementation, we can use some existing tools for these computations including point multiplication, bilinear pairing evaluation, and hash function evaluation over elliptic curves. The protocol is based on the elliptic curve cryptography (ECC) and thus it has high security complexity with short key size.

11 Implimentation Issues

ECC requires the use of two types of mathematics:

- Elliptic curve point arithmetic
- The underlying finite field arithmetic.

For implimentation of ECC based protocol, we have to select the underlying finite field. An elliptic curve is a set of points specified by two variables that are elements over a field. A field is a set of elements with two custom-defined arithmetic operations, usually addition and multiplication.) Most of the computation for ECC takes place at the finite field level. The two most common choices for the underlying finite field are:

- $\mathbb{F}_{2^m}$, also known as characteristic two or even (containing 2m elements, where m is an integer greater than one)
- $\mathbb{F}_p$, also known as integers modulo p, odd, or odd prime (containing p elements, where p is an odd prime number).

12 Conclusion

The security of the proposed protocol is based on difficulty of breaking the Elliptic Curve Diffie-Hellman problem and one way hash function. It archives deniable authentication as well as confidentiality. Also it is resistant against Man-in-Middle attack. It is an non-interactive protocol. The attractiveness of ECC will increase relative to other public-key cryptosystems as computing power improvements force a general increase in the key size. The benefits of this higher-strength per-bit include higher speeds, lower power consumption, bandwidth savings, storage efficiencies, and smaller certificates. Therefore it can be easy to implemented in mobile devices such as PDA, smart card etc. Since the protocol is based on the elliptic curve cryptography (ECC) and thus it has high security complexity with short key size.

References

1. Dwork, C., Naor, M., Sahai, A.: Concurrent zero-knowledge. In: Proc. 30th ACM STOC 1998, Dallas TX, USA, pp. 409–418 (1998)
2. Kar, J.P., Majhi, B.: A Novel Deniable Authentication Protocol based on Diffie-Hellman Algorithm using Pairing techniques. In: ACM International Conference on Communication, Computing and Security (ICCCS 2011), NIT, Rourkela, India, pp. 493–498 (2011)

3. Fan, L., Xu, C.X., Li, J.H.: Deniable authentication protocol based on Diffie-Hellman algorithm. Electronics Letters 38(4), 705–706 (2002)
4. Jiang, S.Q.: Deniable Authentication on the Internet, Cryptology ePrint Archive: Report (082) (2007)
5. Koblitz, N.: A course in Number Theory and Cryptography, 2nd edn. Springer (1994)
6. Menezes, A., Van Oorschot, P.C., Vanstone, S.A.: Handbook of applied cryptography. CRC Press (1997)
7. Bellar, M., Rogaway, P.: Random oracles are practical: a paradigm for designing efficient protocols. In: Proceedings of the 1st CSS, pp. 62–73 (1993)
8. Hankerson, D., Menezes, A., Vanstone, S.: Guide to Elliptic Curve Cryptography. Springer (2004)
9. Certicom ECC Challenge and The Elliptic Curve Cryptosystem, http://www.certicom.com/index.php.
10. Dwork, C., Naor, M., Sahai, A.: Concurrent zero-knowledge. In: Proceedings of 30th ACM STOC 1998, pp. 409–418 (1998)
11. Aumann, Y., Rabin, M.O.: Authentication, Enhanced Security and Error Correcting Codes. In: Krawczyk, H. (ed.) CRYPTO 1998. LNCS, vol. 1462, pp. 299–303. Springer, Heidelberg (1998)
12. Diffie, W., Hellman, M.E.: Directions in cryptography. IEEE Transactions on Information Theory 22, 644–654 (1976)
13. Shi, Y., Li, J.: Identity-based deniable authentication protocol. Electronics Letters 41, 241–242 (2005)
14. Shoup, V.: Sequences of games: a tool for taming complexity in security proofs, in Cryptology ePrint Archive: Report 2004/332, http://eprint.iacr.org/2004/332
15. Deng, X., Lee, C.H., Zhu, H.: Deniable authentication protocols. IEE Proceedings. Computers and Digital Techniques 148, 101–104 (2001)

Mobile Security with Location-Aware Role-Based Access Control

Nils Ulltveit-Moe and Vladimir Oleshchuk

University of Agder, Service Box 509, 4898 Grimstad, Norway
{nils.ulltveit-moe,vladimir.oleshchuk}@uia.no

Abstract. This paper describes how location-aware Role-Based Access Control (RBAC) can be implemented on top of the Geographically eXtensible Access Control Markup Language (GeoXACML). It furthermore sketches how spatial separation of duty constraints (both static and dynamic) can be implemented using GeoXACML on top of the XACML RBAC profile. The solution uses physical addressing of geographical locations which facilitates easy deployment of authorisation profiles to the mobile device. Location-aware RBAC can be used to implement location dependent access control and also other security enhancing solutions on mobile devices, like location dependent device locking, firewall, intrusion prevention or payment anti-fraud systems.

Keywords: location-aware RBAC, GeoXACML, mobile security.

1 Introduction

The objective of this paper is to investigate how location-aware RBAC policies can be implemented in the eXtensible Access Control Markup Language (XACML), which is an authorisation policy language [15]. The solution is based on existing profiles for RBAC [3], and the Geospatial eXtensible Access Control Markup Language (GeoXACML) for location based access control [2].

There are numerous application possibilities for location-aware RBAC. It can be useful for automatic locking/unlocking of the phone based on location. The phone can for example be automatically unlocked in the work premises and at home, but not anywhere else. Another possibility is to use location-aware RBAC for mobile payment applications, to handle location-dependent threats for a mobile payment service. Payment may for example not be recommended or permitted in certain areas. This can either be due to threats against the mobile terminal, like the threat of physical theft or the risk of a localised cyber attack against the mobile terminal in a given location, for example at a rogue access point or bluetooth attacks. It can also be because the risk of fraud is considered large in the given location, based on past known incidents. Another strategy is to explicitly allow mobile payments by letting the user authorise a new location by providing his credentials. This can be used to reduce the risk of fraud, since

R. Prasad et al. (Eds.): MOBISEC 2011, LNICST 94, pp. 172–183, 2012.

mobile payments then only would work in locations and with vendors that were explicitly authorised by the owner.

We assume that the exchange of information related to XACML policy administration can be performed securely, for example over an encrypted link with signed messages, to ensure the confidentiality and integrity of the XACML policy management. It is furthermore assumed that the storage and execution environment can be secured using trusted computing or similar techniques. The paper does not go into details on the authentication process, which can be covered using existing methods and protocols, for example based on the Security Assertion Markup Language (SAML).

This paper is organised as follows: The next section describes how the RBAC profile of XACML implements role-based access control. Section 3 introduces GeoXACML, which is an extension of XACML that provides support for fine-grained authorisation based on geographical data types and functions. This section furthermore elaborates on how location-aware RBAC can be implemented using GeoXACML. Section 4 gives an example of a role-based authorisation policy for intrusion prevention systems based on GeoXACML. Section 5 discusses advantages and disadvantages with the proposed solution, Section 6 goes through related work and Section 7 concludes the paper and discusses future work.

2   RBAC in XACML

Subsequent sections assume that the reader has a basic understanding of XACML. In the following, the XACML 3.0 namespace is denoted as *&xacml;*, the XML Schema namespace is denoted as *&xs;* and our own extensions are defined in the namespace *http://www.prile.org:*, denoted by *&prile;* and roles, defined as *&xacml;:subject:role:*, are in short denoted as *&role;*. The GeoXACML namespace *urn:ogc:def:dataType:geoxacml:1.0:* is in short denoted *&geox;*.

The Organization for the Advancement of Structured Information Standards (OASIS) has defined core and hierarchical RBAC profiles of XACML 2.0 [3]. Figure 1 illustrates how the XACML RBAC profile can be implemented in a Service Oriented Architecture. The most noticeable difference, compared to the standard RBAC model, is that the authorisation model is subdivided into three main functions:

- a *Role Enabling Authority (REA)* that is responsible for managing the User Assignment (UA) mapping (users to roles) in the standard RBAC model;
- an *Identity Manager* that both authenticates the users towards different roles, manages user sessions and can send authorisation requests to access given objects or resources towards the XACML Policy Decision Point (PDP);
- the *XACML RBAC profile* that gives authenticated sessions, with a set of enabled roles, access to given objects/resources based on the permissions these roles have in the XACML PDP.

The REA can be combined with the Identity Manager into an Identity Provider or Single Sign-On (SSO) service that provides authentication and authorisation

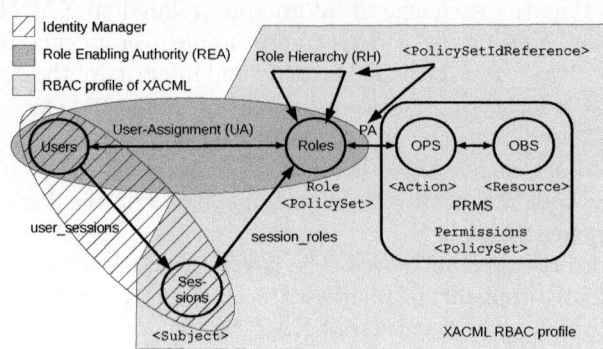

Fig. 1. Implementation of the RBAC model in a Service Oriented Architecture (SOA), based on the RBAC profile of XACML

of users. Examples of such services can be LDAP directory services or federated services like Shibboleth[1] that can use SAML in web services to perform authorisation requests towards the PDP running the XACML RBAC profile. SAML assertions can then be verified by the PEP, presuming that the PEP is trusted. This paper does however not go into the details of the SAML-XACML interaction for space reasons.

The XACML RBAC profile [3] defines how XACML policies can be used to implement the core and hierarchical parts of the NIST RBAC model [9]. This profile expresses:

- *RBAC user_sessions* that are implemented as XACML *<Attributes>* representing authorised users and currently enabled roles for each user in the XACML *<Subject>* elements;
- *RBAC Roles* are expressed as Role *<PolicySet>* elements;
- *RBAC Objects (OBS)* are expressed using XACML *<Resource>* elements;
- *RBAC Operations (OPS)* are expressed using XACML *<Action>* elements;
- *RBAC Permissions (PRMS)* are expressed using XACML Permission *<PolicySet>* elements;
- *RBAC Permissions Assignment (PA)* and *Hierarchical RBAC* are implemented using the *<PolicySetIdReference>* element which refers to other Permission *<PolicySet>* elements.

The XACML Role and Permission policy sets make it easy to extend the permission assignments, to handle location dependent roles or permissions, by adding additional constraints on the respective policy sets. Permissions from different Permission policy sets can be aggregated by including a *<PolicySetIdReference>*

---

[1] Shibboleth can be found at http://shibboleth.internet2.edu

element to each set of permissions that should be included into the policy set
of a given role. The role hierarchy is therefore implicitly defined in the XACML
RBAC profile by adding *<PolicySetIdReference>* pointers to inherited permissions from other roles, as shown in Fig. 2.

A difference between the XACML RBAC profile and the standard NIST
RBAC model, is that the definition of RBAC *Users* and *Sessions* normally are
external to the XACML RBAC profile. The RBAC XACML profile presumes
that the role(s) and session specified in the XACML request are valid for the
given user. Handling of which roles a given user is allowed to enable must instead
be done in the REA and sessions are authenticated by the Identity Manager.

An additional restriction in the XACML RBAC profile, is that an RBAC
compliant XACML PDP must ensure that Permission *<PolicySet>* instances
can never be used as the initial policy of an XACML PDP [3]. This can for
example be ensured by keeping Role and Permission policy sets in separate
PDP's or by adding a semantic restriction that only Role policy sets can be
accessed directly in requests towards the Policy Enforcement Point (PEP).

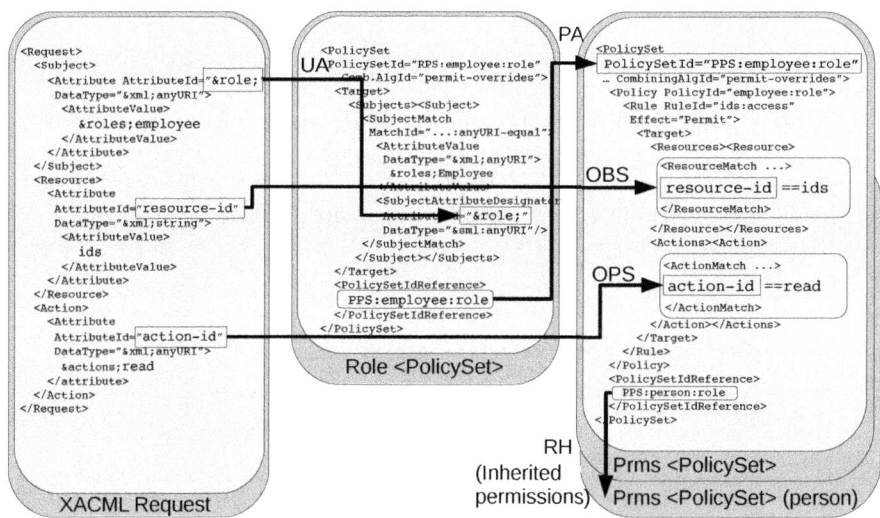

Fig. 2. Inter-dependencies between the XACML Request, the Role policy set and the
Permissions (Prms) policy set both for core and hierarchical RBAC

The set of *<Subject>* attributes in the XACML request furthermore contains
the *role* or *a set of roles* being enabled by this session. The *<Resource>* attribute
of the request refers to the object(s) (OBS) being authorised, and the *<Action>*
attribute of the request describes the operation(s) (OPS) that will be performed
on the object(s). Fig. 2 shows that the permission assignments (PA) is done by

adding a *<PolicySetIdReference>* pointer to the Permissions *<PolicySet>* that contains the permissions for the given role.

The approach chosen for the XACML RBAC profile will in general provide a lower granularity for PA than the traditional RBAC model, since it refers to a *<PolicySet>* for a given role, instead of providing a true many to many relationship between permissions and roles. The XACML RBAC profile does in other words implement roles, but not strictly according to the NIST RBAC standard. It is perhaps neither viable nor desirable to implement the RBAC standard with finer PA granularity, since the current design allows for some flexibility in the permission handling that does not exist in standard RBAC. It is for example easy to add support for location handling or other constraints on permission level. It may on the other hand not be viable to implement very fine-grained permission handling due to the XML parsing overhead in XACML. This means that the XACML RBAC model is a hybrid between XACML and RBAC.

3   GeoXACML

GeoXACML defines an extension of XACML for spatial data types and spatial authorisation functions. The spatial data types are based on the Geographical Markup Language version 3 (GML3) [14]. These data types and functions can be used to define spatial constraints for XACML based policies [2], which means that it is possible to support declaration and enforcement of access restrictions on geographic information. GeoXACML defines mainly the geometry model for geometric data types in access rules and geometric functions that can operate on these geometric data types.

```
<Condition FunctionId="&geox;geometry-intersects">
  <Apply FunctionId="&geox;geometry-one-and-only">
    <SubjectAttributeDesignator AttributeId="my-position"
      DataType="&geox;geometry"/>
  </Apply>
  <AttributeValue DataType="&geox;geometry">
    <gml:Polygon id="Grimstad" srsName="urn:ogc:def:crs:EPSG:6.6:4326"
    xmlns:gml="http://www.opengis.net/gml">
      <gml:exterior>
        <gml:LinearRing>
          <gml:posList dimension="2">
            58.34 8.59 58.5 8.60 ...
          </gml:posList>
        </gml:LinearRing>
      </gml:exterior>
    </gml:Polygon>
  </AttributeValue>
</Condition>
```

Fig. 3. GeoXACML geographical *Condition* example

3.1   Implementing Location-Aware RBAC Based on GeoXACML

Location-aware RBAC can be implemented by adding constraints in the form of *Conditions* on the role and permission policy sets that evaluate a logical expression based on GeoXACML geographical functions. Fig. 3 shows an example XACML condition that checks whether the device position reported by the *<Subject>* attribute *my-position* intersects with or is contained in the polygon defining the town *Grimstad*. This subject attribute contains the current GPS position of the device, for example represented as a GeoXACML circular polygon with radius equal to the measurement uncertainty. This subject attribute can either be provided by the XACML Policy Information Point (PIP) or it can be passed in as part of the XACML request. The function *geometry-one-and-only* is needed because the *<SubjectAttributeDesignator>* element returns a bag of potentially zero or more elements. This function returns the first element of type geometry, and if there are no such elements or more than one element, then the function returns *INDETERMINATE*.

Adding location dependent constraints on the Permissions *<PolicySet>* gives the possibility to place additional geographical restrictions on inherited permissions. This effectively means that it is possible to add geographical restrictions on any level desired, from a course-grained role level and to a fine grained permissions level. XACML Constraints also make it easy to add permissions that apply everywhere but a given set of locations. For example a firewall rule that is applicable for any position p except an area A can be expressed by inverting the result of the geometrical inclusion test, i.e. $p \notin A$.

It should also be noted that the proposed model also supports temporally aware RBAC policies based on XACML. This can be implemented by adding XACML *<Condition>* constraints based on standard XACML functions for time handling in the Role or Permission policysets.

3.2   Spatial Separation of Duties Constraints in GeoXACML

This section sketches how spatial separation of duties constraints can be implemented in GeoXACML. A deficiency with the OASIS XACML RBAC profile is that it does not support static or dynamic separation of duties (SSD/DSD), as specified in the RBAC standard [3]. The reason for this, is that the REA and Identity Management functions are defined outside the XACML RBAC profile, which means that the assignment of users to roles and management of active sessions are not defined as part of the XACML RBAC profile. This also means that a basic assumption is that the XACML authorisation function must trust the REA in order to implement separation of duties constraints.

SSD constraints mean that a user must not be authorised with conflicting roles [9]. These constraints can be implemented in XACML if the REA exposes the mapping of currently authorised users to roles as XACML attributes via the PIP. This can for example be implemented using a SAML *AttributeQuery* for a given user, which implies that the user name must be passed in as part of the XACML request in order to enforce SSD. XACML *<Condition>* constraints on the user-role mapping can then be used in the Role *<PolicySet>* to verify static separation of duty constraints by counting the number of enabled roles that are in the set of conflicting roles in the REA for a given user.

DSD puts restrictions on which roles that can be enabled at the same time in the same user session or across different user sessions [9]. DSD can in a similar way be implemented in XACML if the REA exposes the mapping of active user sessions to roles as XACML attributes via the PIP. This means that XACML *<Condition>* constraints on the user session resources from the Identity Manager can be used in the Role *<PolicySet>* to check for conflicting roles within or across user sessions. This will require that a unique session identifier, for example the SAML session ID, is sent as an attribute of the XACML request.

A central lock manager, for example as suggested in [7], will then be required to ensure safe policy checks for concurrent authorisation requests, so that the SSD/DSD constraints ensure that less than a given number of sessions for a given role are active at the same time. This central lock manager needs to be extended to block updates of sessions and their roles by the REA and the Identity Manager during checking of SSD/DSD constraints to avoid race conditions that could violate the constraints.

Spatial SSD constraints (SSSD) furthermore imply that if a user is assigned to a role in one location, the user cannot be assigned to another role in this location if these two roles are conflicting [10]. In a similar way, Spatial DSD constraints (SDSD) ensure that constraints on which roles that can be enabled in the same user session or across user sessions can be enforced for a given location. The above mentioned scheme can easily be extended to support SSSD or SDSD by adding GeoXACML geographical restrictions to the respective SSD or DSD XACML *<Condition>* constraints on the Role *<PolicySet>*.

Implementing dynamic separation of duties constraints for the XACML RBAC profile is not trivial, since it will require both a central lock manager and an extension of the REA in order to expose the mapping of active user sessions to roles. It is also problematic from both security and privacy perspectives that XACML has access to all active sessions for all users, especially if this occurs in a federated environment where the authenticated sessions may belong to other authorisation functions.

4   IPS Policy Example

This section discusses location based permissions for controlling Intrusion Prevention System (IPS) or firewall rules which can be used to provide more flexible protection of mobile terminals. This can be implemented by adding a separate XACML *Action*, for example *configure-ids* or *configure-firewall*, that the IPS or firewall can use to authorise and update their rule sets for a given set of user roles. Some IPS or firewall rules will then be enabled at any position. Other rules may be enabled or disabled in certain locations or for certain roles or combination of roles. This can be done by having a separate threat manager service that

```
<PolicySet PolicySetId="PPS:payment:role"
  PolicyCombiningAlgId="&xacml;policy-combining-algorithm:permit-overrides">
  <Policy PolicyId="IPS:bluetooth:permissions"
    RuleCombiningAlgId="&xacml;rule-combining-algorithm:permit-overrides">
    <Rule RuleId="IPS:payment:permissions:in:Grimstad"
      Effect="Permit">
      <Target>
        <Resources><Resource><AnyResource/></Resource></Resources>
        <Actions><Action>
          <ActionMatch MatchId="&xacml;function:string-equal">
            <AttributeValue DataType="&xs;string">
              configure-ips
            </AttributeValue>
            <ActionAttributeDesignator AttributeId="&xacml;action"
              DataType="&xs;string"/>
          </ActionMatch>
        </Action></Actions>
      </Target>
      <Condition FunctionId="&geox;geometry-intersects">
        <Apply FunctionId="&geox;geometry-one-and-only">
          <SubjectAttributeDesignator AttributeId="my-position"
            DataType="&geox;geometry"/>
        </Apply>
        <AttributeValue DataType="&geox;geometry">
          <gml:Polygon id="Grimstad" srsName="urn:ogc:def:crs:EPSG:6.6:4326"
          xmlns:gml="http://www.opengis.net/gml">
            <gml:exterior><gml:LinearRing>
              <gml:posList dimension="2">
              58.34 8.59 58.5 8.60 ...
              </gml:posList>
            </gml:LinearRing></gml:exterior>
          </gml:Polygon>
        </AttributeValue>
      </Condition>
    </Rule>
    <Obligations>
      <Obligation ObligationId="&prile;ips-rule:apply"
        FulfillOn="Permit">
        <AttributeAssignment AttributeId="&prile;enable:snort:rule:1"
          DataType="&prile;snortrule">
          reject tcp $HOME_NET any -&gt; $HTTP_SERVERS $HTTP_PORTS (\
          msg:"Mobile␣payment␣attack␣rejected␣(fraud␣risk)";
          flow:to_server,established; uricontent:"http://www.mybank.com";
          nocase;  classtype:attempted-recon; sid:10000001; rev:1;)
        </AttributeAssignment>
      </Obligation>
    </Obligations>
  </Policy>
</PolicySet>
```

Fig. 4. Example *Permissions* *<PolicySet>* for location-aware IDS policy

enables or disables the rules based on location dependent threats. Such a service checks the threat policies and updates IDS and firewall rules continually based on parameters such as time, device speed and threat profile updates.

Fig. 4 shows an example *Permissions* *<PolicySet>* that is used to configure security requirements for the *payment role* of a mobile device. The *<PolicySet>* requires that the IPS should be installed with a rule that rejects connections to *http://www.mybank.com* if the user is within the Grimstad area, for example to reduce the risk of fraud. This rule will be installed the next time the threat management PEP performs an XACML request on the current user session. If the user session has the *payment role* enabled and the PEP issues a *configure-ips* action, then the PDP will reply with a *Permit* decision with an *Obligation* to apply the given IPS rule.

If the user later moves outside of the Grimstad area, then this restriction will be removed the next time the threat management PEP is scheduled.

5 Discussion

A challenge with the proposed approach, is that the policy management in the PAP is quite complex. The role hierarchy is for example implicitly defined via links between permissions policy sets in the XACML RBAC profile, something that is not very intuitive. Routine policy generation tasks can however be automated, and the resulting API can be made similar to the functional specification in traditional RBAC [9], however with the necessary adaptations to handle geographical constraints for roles and permissions and also for managing the permissions policyset. The details of the modified API is however beyond the scope of this paper.

A disadvantage with location dependent firewall rules, is that it may not be very user friendly or intuitive. The final user may for example be puzzled if location based policies prohibit access, since it is not clear to the user *why* access is prohibited. This can however be handled by adding an additional XACML <*Obligation*> to the policy that will specify a *reason* for why access is prohibited, for example to give a notice that "Access is prohibited in this location to service XXX due to risk of fraud".

It is also possible to use other means than firewall or IPS rules to restrict access. It would probably be more useful for a mobile payment application if a successful authorisation returns an obligation containing part of a cryptographic key needed to gain access to the service, rather than managing firewall or IPS rules. In particular on phones where the firewall or IPS run as untrusted applications, since these then easily can be disabled by a determined adversary that has stolen or hacked into and compromised the phone.

An issue that needs to be considered is the reliability of the positioning system, and whether positions easily can be forged[2] in order to avoid location dependent access restrictions. GPS is for instance relatively accurate in an outdoor environment, but it quickly loses connection indoors.

Another problem is that GPS data coming from the device itself is not guaranteed to be correct if the operating system is compromised. For example if a rootkit or rogue GPS device driver intercepts and changes the GPS signals on the fly, in order to influence the authorisation process.

This risk can be reduced to some extent using traditional mitigations like AntiVirus and good patching and update procedures. It is however more efficient to use techniques like cryptographically signed drivers and secure booting of the phone operating system where a Trusted Platform Module (TPM) in future mobile phones ensures that unauthorised modifications, for example by a rootkit,

---

[2] GPS Spoofing Countermeasures http://www.homelandsecurity.org/bulletin/Dual%20Benefit/warner_gps_spoofing.html

causes the TPM to refuse booting the phone [8]. The risk of being infected by a rootkit can probably not be eliminated completely, since there always will be a risk of stack or buffer overflow vulnerabilities or other flaws also in trusted and signed applications.

This means that using a positioning technique like GPS has its deficiencies when used for access control purposes. These deficiencies will also apply if the access control is used to manage location dependent threats. It is in some cases hard to put a clear distinction between areas that are safe and areas that are unsafe and if a safety margin is added to location threats, then this may harm legal businesses which clearly is undesirable. Using near-field communications (e.g. RFID) together with a *proof-of-location* protocol is a promising technology that can be used as an additions means to verify known locations [12]. This can improve the precision of location-aware RBAC and also reduce the risk of harming legal businesses.

6   Related Work

This paper describes how location-aware role based access control can be implemented by combining GeoXACML and the RBAC profile of XACML. Such a GeoXACML-based RBAC solution has to the best of our knowledge not been published before.

There are numerous previous works on location-based or spatial RBAC models. This discussion does not cover all RBAC models, but compares our solution to some of the well known traditional location based RBAC solutions. SRBAC is one of the earlier models of a spatial RBAC system [10, 11], and it has been suggested to define an XACML-based location aware RBAC model based on SRBAC [1]. Our model has moved the location constraints into the Role and Permissions (PRMS) definitions instead of expressing location references as an explicit relationship between *Roles-Locations* and *Locations-Operations* as SRBAC does.

GEO-RBAC extends the traditional RBAC model with spatial entities that are used to model objects, user positions and geographically bounded roles [4, 6]. A model for enforcing spatial constraints for mobile RBAC systems based on GEO-RBAC is described in [12]. Roles in GEO-RBAC are enabled based on the position of the user. GEO-RBAC does however not support defining location on object permissions. LRBAC and LoT-RBAC solve this deficiency [13, 5].

The nice feature with all native RBAC models, is that they are fast, efficient and have well defined formal definitions founded in set theory. Extensibility for these models has however evolved over time, and is not yet standardised for location or time based RBAC models. XACML is on the other hand designed from the start to be extensible, with rich semantics for defining constraints. The RBAC profile of XACML implements role based access control, however not with the same level of granularity as native RBAC solutions and also with deficiencies when it comes to enforcing separation of duties constraints. The extensibility of XACML makes it easy to embed and combine other XML-based standards, like in this case the RBAC profile of XACML and GeoXACML, and it also makes it trivial to add arbitrary constraints both on roles and permissions. An advantage with our model is that the physical location definition is embedded in the XACML policies, which simplifies deployment of location-based policies. This is useful for outsourced managed security services, where updated threat profiles then can be generated and deployed according to the needs of the mobile terminal.

7   Conclusions

This paper demonstrates how location-aware RBAC can be implemented in GeoXACML, based on the XACML RBAC profile. It furthermore shows how static and dynamic separation of duties constraints can be implemented for this solution. An advantage is that it allows for embedding geographical information directly into the authorisation policies, instead of using logical addressing of locations.

A potential disadvantage with an XACML based RBAC solution, is that it may scale more poorly than a traditional RBAC solution, especially if a large number of roles and permissions are involved. The main problem in XACML will then be the parsing overhead of XML documents. This can to some extent be mitigated using a decision cache under the presumption that location threats do not change too rapidly.

The general discussion also shows that it can be difficult to put a clear distinction between areas that are safe and areas that are unsafe due to the inherent unreliability of positioning systems and also unreliability in the definition of what areas are considered unsafe. This may harm legal businesses which clearly is undesirable. There are in other words several concerns with location-based threat management and location-based authorisation.

Future work includes implementing and testing the location-aware RBAC solution, including to elaborate and demonstrate how well the proposed solution works for static/dynamic spatial separation of duties. There is also more work required on how to reliably identify locations and properly secure the overall solution.

Acknowledgments. This work is funded in part by Telenor Research & Innovation under the contract DR-2009-1.

References

1. Aburahma, M., Stumptner, R.: Modeling location attributes using XACML-RBAC model. In: MoMM 2009, Kuala Lumpur, Malaysia, p. 251 (2009)
2. Matheus, A. (ed.): OGC 07-026r2 Geospatial eXtensible Access Control Markup Language (GeoXACML) version 1.0 (2007),
   http://portal.opengeospatial.org/files/?artifact_id=25218

3. Anderson, A. (ed.): Core and hierarchical role based access control (RBAC) profile of XACML v2.0 (2005),
http://docs.oasis-open.org/xacml/cd-xacml-rbac-profile-01.pdf
4. Bertino, E., Catania, B., Damiani, M.L., Perlasca, P.: GEO-RBAC: a spatially aware RBAC. In: ACM SACMAT, p. 37 (2005)
5. Chandran, S.M., Joshi, J.B.D.: *LoT-RBAC*: A Location and Time-Based RBAC Model. In: Ngu, A.H.H., Kitsuregawa, M., Neuhold, E.J., Chung, J.-Y., Sheng, Q.Z. (eds.) WISE 2005. LNCS, vol. 3806, pp. 361–375. Springer, Heidelberg (2005)
6. Damiani, M.L., Bertino, E., Catania, B., Perlasca, P.: Geo-rbac: A spatially aware rbac. ACM TISSEC 10(1), 2 (2007)
7. Dhankhar, V., Kaushik, S., Wijesekera, D.: XACML Policies for Exclusive Resource Usage. In: Barker, S., Ahn, G.-J. (eds.) Data and Applications Security 2007. LNCS, vol. 4602, pp. 275–290. Springer, Heidelberg (2007)
8. Dietrich, K., Winter, J.: Implementation Aspects of Mobile and Embedded Trusted Computing. In: Chen, L., Mitchell, C.J., Martin, A. (eds.) Trust 2009. LNCS, vol. 5471, pp. 29–44. Springer, Heidelberg (2009)
9. Ferraiolo, D.F., Sandhu, R., Gavrila, S., Kuch, D.R., Chandraouli, R.: Proposed NIST Standard for Role-Based Access Control (2001)
10. Hansen, F., Oleshchuk, V.: Spatial role-based access control model for wireless networks. In: IEEE 58th VTC, pp. 2093–2097 (2003)
11. Hansen, F., Oleshchuk, V.: SRBAC: A spatial role-based access control model for mobile systems. In: NORDSEC, pp. 129–141 (2003)
12. Kirkpatrick, M.S., Bertino, E.: Enforcing spatial constraints for mobile RBAC systems. In: ACM SACMAT, pp. 99–108. ACM (2010)
13. Ray, I., Kumar, M., Yu, L.: LRBAC: A location-aware role-based access control model. In: Information Systems Security, pp. 147–161 (2006)
14. Cox, S., et al. (eds.): OGC 02-023r4 OpenGIS Geography Markup Language (GML) Encoding Specification Version 3.00 (2002),
https://portal.opengeospatial.org/files/?artifact_id=7174
15. Moses, T. (ed.): OASIS eXtensible Access Control Markup Language (XACML) Version 2.0 (2005), http://docs.oasis-open.org/xacml/2.0/access_control-xacml-2.0-core-spec-os.pdf

Author Index

GPSR Compliance

*The European Union's (EU) General Product Safety Regulation (GPSR)
is a set of rules that requires consumer products to be safe and our
obligations to ensure this.*

*If you have any concerns about our products, you can contact us on
ProductSafety@springernature.com*

In case Publisher is established outside the EU, the EU authorized
representative is:

Springer Nature Customer Service Center GmbH
Europaplatz 3
69115 Heidelberg, Germany

Batch number: 09490862

Printed by Printforce, the Netherlands